Co-ordinated SCIENCE

Chemistry

Teacher's guide

Brian Beckett   RoseMarie Gallagher
Paul Ingram   Stephen Pople

Oxford University Press

Oxford University Press
Walton Street
Oxford OX2 6DP

Oxford New York Toronto
Delhi Bombay Calcutta Madras Karachi
Petaling Jaya Singapore Hong Kong Tokyo
Nairobi Dar es Salaam Cape Town
Melbourne Auckland
and associated companies in
Berlin Ibadan

Illustrated by Ann and Andrew Miller

Typeset by Microset Graphics Ltd, Basingstoke

Produced by AMR for Oxford University Press

ISBN 019 914314 5

Printed in Great Britain by Ebenezer Baylis & Son Ltd, Worcester

Contents

Introduction

Co-ordinated Science from Oxford will help you to provide a broad and balanced programme of study leading to GCSE Science. There is a *Student's Book,* an *Activities Book* and a *Teacher's Guide* for each of Physics, Chemistry and Biology.

There are three sections in this guide:

- For each double-page spread in the *Chemistry Student's Book* there is
  - a concise set of notes drawing the conceptual links between the separate science subjects. These can be used to minimize overlap between subjects.
  - cross-referencing to related sections in the other *Student's Books* to improve awareness of the ideas that students may be bringing to a particular topic.
  - a list of relevant activities in the *Activities Books* and *Teacher's Guides.*

- A bank of photocopiable activity sheets for use at school or home which present a range of challenges, drawing on knowledge across the sciences.

- A set of photocopiable practicals presented in a form to help structure assessment. A basic instruction sheet for students is augmented by optional cue sheets. There is full guidance for the teacher and technician on equipment needed and assessment criteria.

Together with the *Student's* and *Activities Books* this guide will help you to co-ordinate your GCSE balanced science teaching programme.

TEACHER'S NOTES

1 Particles

1.1   Everything is made of particles

Topics covered: particles in matter; evidence for particles — diffusion and Brownian motion

Brownian motion and diffusion both provide evidence to suggest that matter is made up of tiny particles. Diffusion is the means by which materials move in and out of living cells.

Teachers may wish to draw comparisons between the larger 'particles' (cells) from which living organisms are built, and the much smaller particles of which all materials are made. (Cells, which are of course made up of atoms, are of the order 10^8 times larger than atoms.)

Students

Biology
2.1   Cells
2.3   Diffusion
E2   Size of cells
E5.A   Diffusion in microscopic organisms

Physics
2.10   Molecules, Brownian motion and diffusion
5.1   Atoms

Activities

Chemistry
Looking at diffusion (p. 14)
To dissolve or not to dissolve (p. 52)

Biology
Diffusion (p. 10)
Osmosis (p. 12)
Patterns of movement (**TG** Activity 8)

1.2   Solids, liquids and gases
1.3   The particles in solids, liquids and gases

Topics covered: characteristics of solids, liquids and gases; melting, evaporation and boiling; melting and boiling points; particles (molecules) in solids, liquids and gases; condensing and solidifying

Heat can change a solid into a liquid or a liquid into a gas. Adding impurities to a material may change its melting and boiling points.

Evaporation produces cooling. The effect is used to cool the body. It is also used to produce cooling in a 'fridge.

In the water cycle, water leaves the ground by evaporation and returns to it by condensation.

Students

Biology
5.7   Evaporation and temperature control
9.5   The water cycle

Physics
2.10   Molecules in solids, liquids and gases
2.22   Melting and freezing
2.23   Evaporation and boiling
2.24   Cooling by evaporation

Activities

Biology
Transpiration (p. 24)

Physics
Finding the melting point of wax (p. 20)

1.4   A closer look at gases

Topics covered: change of gas pressure with temperature; change of pressure with volume; change of volume with temperature; diffusion of gases

Pressure is defined as force/area. There are mathematical relationships between the pressure, volume and temperature of a fixed mass of gas.

Diffusion, both gaseous and liquid , is an essential part of many biological processes. Materials pass in and out of living cells by diffusion. It is the mechanism by which plants absorb water, the intestines absorb food, and gases pass in and out of the bloodstream. Osmosis is a special kind of diffusion in which only the smaller molecules are able to pass through a membrane.

1.5	**Mixtures**

Topics covered: mixtures; solutions and solvents; suspensions

The cytoplasm of a cell is a mixture of many materials. Blood is a mixture, so are the foods taken into the body. Air is a mixture of gases, often including many pollutants. One pollutant, cigarette smoke, is a mixture of over 1000 chemicals.

Elements are a mixture of different varieties of atom, called isotopes. In nuclear fuel, the natural isotopic mixture of uranium is enriched with extra uranium-235.

1.6	**Solubility**

Topics covered: saturated solutions; solubility in water; change of solubility with temperature; solubility of gases

The physical properties of a liquid may be affected if materials are dissolved in it. For example, salt and antifreeze both lower the freezing point of water.

Water moves from one cell to another by osmosis. The strengths of the solutions in the cells determines which direction the water moves.

Fish use their gills to extract the oxygen dissolved in water. In animals and plants, materials are transported in solution, with water as the solvent. Pollutants dissolved in water can readily enter food chains and threaten life.

Tastes and smells are detected when chemicals dissolve in the liquids which coat the nose lining and the tongue.

Activities

Chemistry
To dissolve or not to dissolve (p. 52)
Investigating precipitates (**TG** Assessment 3)

Biology
A fishy problem (**TG** Activity 6)

1.7	**Separating mixtures (I)**
1.8	**Separating mixtures (II)**

Topics covered: methods of separating mixtures, including filtering and chromatography

In the body, the kidneys filter out waste substances from the blood.

Filtering is an essential feature of water treatment plants and sewage works.

The major constituents of soil can be separated out by allowing a soil/water mixture to settle.

Magnets can be used to separate magnetic and non-magnetic materials. Electromagnets are used for this during the recycling of scrap metal.

Complex chemical techniques have to be used to separate the reusable and non-reusable materials present in nuclear waste.

Air pollution can be reduced by fitting filters to power stations, and by removing harmful chemicals from coal, oil and petrol before they are burnt.

Activities

Chemistry
Separating salt from sand (p. 9)
Separating the colours in ink (p. 10)
Distilling cola (p. 11)
Making Epsom salts (p. 23)
Investigating the limestone cycle (p. 26)
Making a fertilizer (p. 37)

Biology
How to compare soil samples (p. 52)

2 What are these particles?

<table>
<tr><td>2.1</td><td>Atoms, elements and compounds</td></tr>
</table>

Topics covered: size of atoms; elements; metals and non-metals; compounds

Atoms are far too small for their size to be measured directly in most school laboratories, but an estimate can be made of the size of a molecule.

Many properties distinguish elements from each other, including their density, their heat and electrical conductivities and their response to tension and compression forces.

Living organisms are mainly made from compounds of carbon, hydrogen, oxygen and nitrogen. But additional elements are present. These include iron in blood and calcium and phosphorus in bones.

Students

Physics
- 1.2 Density
- 1.3 Measuring density
- 2.10 Measuring the size of a molecule
- 2.18 Good and bad conductors of heat
- 3.4 and 3.5 Refraction and colour
- 4.2 Good and bad conductors of electricity
- 4.3 Stretching and compressing forces

Biology
- 1.2 Living organisms need minerals
- 2.6 Chromosomes and genes
- 3.1 Materials for making plants
- E3 Importance of minerals in plants
- 6.1 Food for body building; minerals in food
- 9.2 The importance of carbon
- 9.3 The importance of nitrogen
- 10.1 Viruses

Activities

Chemistry
Looking at elements (p. 15)
Testing for electrical conductivity (p. 26)

Physics
Finding the melting point of wax (p. 20)

<table>
<tr><td>2.2</td><td>More about the atom</td></tr>
<tr><td>2.3</td><td>Some different atoms</td></tr>
</table>

Topics covered: structure of the atom; atomic number (proton number) and mass number (nucleon number); structures of different atoms; isotopes

Atoms of the hundred or so basic elements are made up of electrons, protons and neutrons. It is the 'loose' outer electrons which make metals good conductors of electricity.

The atoms of each element can exist in a number of different versions, called isotopes. Some isotopes are unstable. When their atoms decay, they change into atoms of a different element.

Students

Physics
- 4.1 Charges from the atom
- 4.2 Moving electrons; conductors
- 5.1 Structure of atoms; isotopes
- 5.5 Radioactive decay; decay products

Activities

Chemistry
Looking at elements (p. 15)
Testing for electrical conductivity (p. 28)

<table>
<tr><td>2.4</td><td>The Periodic Table</td></tr>
<tr><td>2.5</td><td>Some groups in the Periodic Table</td></tr>
</table>

Topics covered: arranging elements in groups; the Periodic Table; alkali metals, halogens, noble gases

Metals have thermal and electrical properties which separate them from most non-metals. The separation of metals from non-metals is also apparent in the Periodic Table.

Teachers may wish to draw comparisons between the principles used to classify elements and those used to classify living organisms.

Students

Physics
2.18   Metals as good conductors of heat
4.2   Metals as good conductors of electricity

Biology
1.3   Classifying living organisms
1.4   Groups of living organisms
E1.B   Naming and classifying organisms

Activities

Chemistry
Looking at elements (p. 15)

2.6   Radioactivity

Topics covered: radioactivity; nuclear radiation and its uses and dangers; radioactive decay and half-life

Nuclear radiation causes ionization. It is dangerous because it damages or destroys living cells by disrupting the functioning of the cell nucleus. It causes cancer, but can also be used to destroy tumours. Natural radiation is partly responsible for the mutations which are a feature of the evolutionary process.

Nuclear radiation has many uses. Some of these are controversial: for example, the irradiation of food to extend shelf-life.

Students

Physics
5.2   Types of nuclear radiation; ionization and its effects
5.3   Treating food with radiation
5.4   Detecting nuclear radiation; uses of radiation
5.5   Radioactive decay; dangerous decay products

Biology
2.1   Inside a cell
2.9   Evolution
9.7   Dangers of radiation
E10.B   Causes and treatment of cancer

Activities

Physics
Radioactivity (**TG** Activity 9)

2.7   Power from the nucleus

Topics covered: fission; chain reactions; nuclear power

Nuclear power is the most controversial of the power generation sources because it is perceived by the public as the most dangerous. The principal concern is the possibility of leakage of radioactive materials from the reactors themselves or from the containers used to transport and store their waste products. Any radioactive materials which leak out eventually enter the food chain through the soil or the water supply. Radioactive materials taken into the body with food or water may become concentrated in the bones or in organs such as the liver and the thyroid gland. The long-term effects can be lethal.

Geothermal energy is a form of nuclear energy. The heat results from radioactive decay in the rocks.

Students

Physics
2.8   Energy resources; alternative sources of energy
4.23   Environmental problems of power generation
5.2   Transporting nuclear waste; dangers of ionization
5.3   Nuclear disasters; dumping nuclear waste
5.3   Radioactivity makes rocks hot; decay hazards
5.6   Nuclear power − fission and nuclear reactors

Biology
6.7   The liver
7.7   The thyroid gland
9.1   Food chains
9.4   Soil
9.5   The water cycle
9.6   Water pollution
9.7   Air pollution
9.8   Geothermal energy
E10.B   Radiation causes cancer

Activities

Chemistry
Energy − what is it? (**TG** Activity 1)

3   Atoms combining

3.1	Why compounds are formed
3.2	The ionic bond
3.3	Some other ions

Topics covered: compounds formed from elements; the noble gases; ions; ionic bonding; ionic compounds; ions of the first 20 elements; transition metal ions; compound ions

Atoms are held together by electric forces; atoms are bonded to each other by electric forces. Electric forces are one component of the family of forces which holds the universe together. Others are magnetic, gravitational and nuclear forces.

In simple kinetic theory, solids, liquids and gases are all made up of particles called molecules. In reality, the picture is more complex. The moving particles may be single atoms, ions or molecules.

Students

Physics
2.1   Gravitational forces
2.10   Molecules and the kinetic theory
4.1   Forces between charges
4.13   Magnetic forces
5.1   Structure of atoms; nuclear forces

Biology
3.1   Protcins and mincrals
3.3   Transport of minerals in plants
6.2   Minerals in food
6.7   Storage and use of minerals

Activities

Chemistry
Burning elements in oxygen (p. 16)
Reacting metals with oxygen (p. 17)
Reacting metals with acids (p. 21)
Making Epsom salts (p. 23)
Making a fertilizer (p. 37)
Light chemistry (p. 62)
Energetic chemicals (**TG** Activity 5)
Investigating precipitates (**TG** Activity 4)
Investigating the electrolysis of compounds
(**TG** Assessment 4)

3.4	The covalent bond

Topics covered: shared electrons; groups of atoms form molecules; covalent compounds

In simple kinetic theory, solids, liquids and gases are all made up of particles called molecules. Strictly speaking, molecules are only those groups of atoms held together by covalent bonding.

Plants take in simple molecules such as carbon dioxide and water. These are converted into the more complex molecules of carbohydrates, fats and proteins. DNA is a complex molecule which stores genetic information in living cells.

Students

Physics
5.7   Balanced forces: model of a covalent bond
5.1   Structure of atoms
2.10   Molecules and the kinetic theory

Biology
2.1   Chemicals in cells
2.6   Molecules of DNA in the cell nucleus
3.1   Substances used to make plants
6.1   Substances in food used for body building

Activities

Chemistry
Investigating carbon dioxide (p. 24)
Investigating the limestone cycle (p. 26)
Cracking oil (p. 44)
Energetic chemicals (**TG** Activity 5)
The hydrogen revolution (**TG** Activity 11)

Biology
Light, carbon dioxide and photosynthesis
(p. 18)

3.5	Ionic and molecular solids
3.6	The metals and carbon

Topics covered: ionic and molecular solids and their properties; the metals and their properties; carbon in the form of diamond and graphite

In simple kinetic theory, solids are made up of particles called molecules. In reality, the particles may be ionic or molecular in nature. Solids occur in a variety of types and, in general, these have different properties. Ionic solids have higher melting and boiling points than molecular solids. Both types of solid are poor conductors of heat and electricity.

The metals and graphite are good conductors of heat and electricity. Graphite is also soft and slippery, which makes it the best material for electrical contacts which slide.

Students

Physics
2.8   Molecules in solids
2.11   Temperatures compared
2.18   Good and bad conductors of heat
2.22   Melting points; molecular structure of ice
2.23   Boiling points
4.2   Good and bad conductors of electricity
4.18   Carbon brushes in electric motors

Biology
2.1   What are cells?

Activities

Chemistry
Separating salt from sand (p. 9)
Looking at elements (p. 15)
Testing for electrical conductivity (p. 28)

4   The mole

4.1   The masses of atoms

Topics covered: relative atomic mass; isotopes; formula mass

The masses of atoms are far too small to be measured conveniently in grams or kilograms. Measurement of relative atomic masses is much more useful. Relative atomic masses are directly related to the number of neutrons and protons in the nucleus.

Students

Physics
2.1   Measuring mass
5.1   Nucleon numbers of different atoms

4.2   The mole

Topics covered: the mole and Avogadro's number

A mole is a number (6.02×10^{23}), much as a dozen, score, century and gross are all numbers. It is possible to obtain a mole of atoms or molecules of a substance by simple mass measurement.

Though it is strictly beyond the scope of GCSE, teachers may wish to point out that if moles of different gases are all at the same pressure and temperature, they also have the same volume.

Students

Physics
1.1   Measuring mass
2.16   Pressure/temperature and volume/temperature relationships for a gas
2.17   Pressure/volume relationship for a gas

4.3   Compounds and solutions

Topics covered: percentages of elements in a compound; concentration of solutions

The concentration of a solution affects its freezing point. Water freezes at a lower temperature if salt or antifreeze is mixed with it.

In plants, water moves from cell to cell by osmosis. The direction of movement depends on the concentrations of the solutions in the cells. The body has complex mechanisms for maintaining the concentrations of the various body fluids at the correct levels.

Pollution levels in water and in air can be expressed as concentrations.

Students

Physics
2.22   Freezing point of salt solution

Biology
2.3   Osmosis and concentrations of solutions
E2   Osmosis and solution concentration
3.3   Osmosis keeps plant cells firm or turgid
E5.B   Regulating the concentrations of solutions in the body
9.6   Water pollution
9.7   Air pollution

4.4   Formulae of compounds

Topics covered: finding formulae from masses; finding masses by experiment

The formulae of compounds can be found by careful measurement of masses.

Students

Physics
1.1   Measuring mass

Activities

Chemistry
Measuring the volume of gas given off in a reaction (**TG** Assessment 1)

5 Chemical change

5.1 | Physical and chemical change

Topics covered: chemical changes; mixtures and compounds; physical changes

During a chemical change, substances react together to form new substances. These may have entirely different properties from the original materials.

Living organisms function through chemical changes. Plants change simple materials into more complex ones during photosynthesis. Animals change digested materials into new body tissue. In a fertilized hen's egg, chemical changes produce a chicken. The sequence of chemical changes in a food chain produces organisms of a variety of different forms.

Examples of physical change include the melting of ice, the evaporation of water, the stretching of springs and the expansion of solids and liquids when heated. Such changes are usually easy to reverse.

Students

Physics
1.16 Stretching and compressing materials
2.13 Expansion of solids
2.15 Expansion of solids
2.22 Melting of ice
2.23 Vaporization of water
4.13 Magnetic materials

Biology
E1.A The body as a chemical factory
2.1 The cell as a chemical factory
3.1 Plants built from simple materials
6.1 New cells built from substances in food
6.5 Food changed into other materials during digestion
6.7 The liver as a chemical factory
6.8 New materials made by microbes
9.1 Food chains
9.2 The carbon cycle
9.3 The nitrogen cycle
9.7 Pollution: physical and chemical effects
E9.A Food chains

Activities

Chemistry
Concrete composition (p. 50)

Physics
Stretching a spring (p. 15)
Finding the melting point of wax (p. 20)
Salt on ice (p. 48)

5.2 | Equations for chemical reactions
5.3 | Calculations from equations

Topics covered: balancing chemical equations; conservation of mass in a chemical change

In a chemical change, the total number of atoms is unchanged. During radioactive decay, the total number of protons + neutrons is unchanged.

Students

Physics
1.1 Units of mass
5.5 Radioactive decay products

Activities

Chemistry
Burning elements in oxygen (p. 16)
Electrolysing water (p. 30)

5.4 | Different types of chemical reaction (1)

Topics covered: Combination or synthesis; decomposition; precipitation; combustion

In combination reactions, two or more substances combine to form one. During decomposition, a single substance breaks down into two or more substances. Such reactions are a feature of many biological processes. Both types of reaction occur in the carbon cycle and in the nitrogen cycle.

Decomposition reactions are used to produce photographs.

The rotting of dead animals and plants is sometimes called decomposition, but the chemical reactions which occur are not necessarily decomposition reactions.

Combustion reactions are the principle means by which energy is obtained for transport, heating, lighting and power. Some of the reaction products pollute the atmosphere.

Students

Physics

1.12	Jet and rocket engines
2.4	Petrol, diesel and human engines
3.7	Light-sensitive film in a camera
4.6	Dangers of electrical fires
4.22	Burning fuels in a power station
4.23	Fuels from unlikely sources

Biology

E1.A	Chemical reactions in the body
2.1	Chemical reactions in cells
3.1	Photosynthesis, plants making food from simple substances
4.4	Chemical changes in respiration
6.1	New cells built from substances in food
6.5	Food changed into other materials during digestion
6.7	The liver as a chemical factory
6.8	New materials made by microbes
7.7	Hormones produce changes
9.1	Food chains; decomposers
9.3	The nitrogen cycle
9.7	Combustion products cause pollution
10.3	Dangerous combustion products of tobacco

Activities

Chemistry
Burning elements in oxygen (p. 16)
Investigating the limestone cycle (p. 26)
Light chemistry (p. 62)
Investigating precipitates (**TG** Assessment 3)

5.5	**Different types of chemical reaction (II)**

Topics covered: oxidation; reduction; reversible reactions

Reversible changes can be chemical or physical. Examples of reversible physical change include changes of state between solid, liquid and gaseous forms of a material, the expansion of solids, liquids and gases, and the magnetization of iron. In contrast, the magnetization of steel is difficult to reverse. And the stretching of a spring or wire is only reversible if the elastic limit isn't exceeded.

Students

Physics

1.16	Stretching materials; the elastic limit
2.6	Machines
2.7	Liquid machines
2.13	Expansion
2.15	Expansion of liquids
2.16	Effect of temperature on pressure and volume of a gas
2.17	Effect of volume on pressure of a gas
2.22	Melting and freezing
2.23	Evaporating and condensing
4.13	Magnetic materials

Biology

9.5	Changes of state in the water cycle

Activities

Chemistry
Burning elements in oxygen (p. 16)
Reacting metals with oxygen (p. 17)
Investigating the limestone cycle (p. 26)

6 Electricity and chemical change

6.1	Conductors and insulators

Topics covered: a simple circuit; testing materials for conduction

The metals and graphite are the only solids which are good conductors of electricity. The greater the resistance of a material, the poorer a conductor it is.

The brain and central nervous system use electrical pulses to monitor and control the body. Nerve fibres conduct electricity – though, unlike wires, the conduction is from the centre to the outside of the fibre, rather than along its length.

Students

Physics
- 4.2 Conductors, semiconductors and insulators; cells and batteries; electrolysis
- 4.3 Current in a simple circuit
- 4.5 Resistance

Biology
- 7.4 Nerve impulses from the inner ear
- 7.5 Nerve cells as conductors
- E7.B Reflex arcs

Activities

Chemistry
Testing for electrical conductivity (p. 28)
Long life (p. 56)
Best battery (p. 58)

Physics
Measuring resistance (p. 28)

6.2	A closer look at electrolysis
6.3	The electrolysis of solutions
6.4	Some uses of electrolysis

Topics covered: electrolysis in detail; the electrolysis of compounds and solutions; uses of electrolysis

Circuits used for electrolysis follow the same basic principles as other electrical circuits.

Electrolysis is used in the production of aluminium and the purification of copper. Pure copper is needed for the wires used in electromagnets, electric motors, electronics and household circuits. In the power industry, aluminium is used for overhead power lines.

Students

Physics
- 4.3 Simple circuit principles
- 4.11 Wires in household circuits
- 4.16 Electromagnets
- 4.18 Electric motors
- 4.22 Power generation; overhead power lines
- 5.14 Transistor circuits

Activities

Chemistry
Testing for electrical conductivity (p. 28)
Electrolysing water (p. 30)
Electroplating (p. 31)
Best battery (p. 58)
Investigating the electrolysis of compounds (**TG** Assessment 4)

7 How fast are reactions?

7.1 | Rates of reaction

Topics covered: what a rate means; rate of a chemical reaction

Rates of change are an important concept in science. Motion can be described in terms of rates of change of distance and of speed. Power is the rate of change of energy. Frequency indicates the rate at which waves are produced. The activity of a radioactive source can be expressed in terms of a count-rate.

The body's power output depends on the respiration and heartbeat rates.

On a world scale, the rate of population growth has caused severe food distribution problems.

Students

Physics
1.4	Speed as rate of change of distance
1.5	Showing rates in graphs
2.5	Power as rate of working
3.8	Frequency of waves
3.15	Frequency of sound waves
4.10	Count-rate and radioactive decay

Biology
E1.A	Rates of reaction in the body
3.4	Transpiration: rate of water loss
4.4	Respiration rate
E4	Respiration rate
5.1	Heartbeat rate
E5.A	Heartbeat rate
6.2	Eating for health :rate of food consumption
8.6	World population growth rate
E8.B	World population growth rate

Activities

Chemistry
Reacting metals with acids (p. 21)

Biology
Photosynthesis: problems to solve (p. 20)

Physics
Measuring acceleration (p. 12)

7.2 | Measuring the rate of a reaction

Topics covered: collecting gas to measure the rate of a reaction

If one product of a chemical reaction is gaseous, the reaction rate can be measured by finding how the volume of the gas increases with time.

The rate of aerobic respiration in the human body depends on the rate of gas exchange in the lungs.

Students

Physics
1.1	Measuring volume

Biology
4.4	Respiration rate
4.5	Gas exchange in the lungs
E4	Respiration rate; gas exchange in the lungs

7.3 | Changing the rate of a reaction (I)
7.4 | Changing the rate of a reaction (II)

Topics covered: factors affecting rates of reaction

The speed of a chemical reaction depends on its temperature. A steady temperature is important for many of the life-maintaining reactions which take place in the body. The body uses vitamins and enzymes in controlling the rates of some of its chemical reactions.

The reactions which cause bacteria to multiply and food to rot all depend on temperature. They can be slowed down by keeping food in a 'fridge.

In plants, the rate of photosynthesis depends on the temperature. It also depends on the light intensity, the carbon dioxide content of the air and the supply of water.

Teachers may wish to draw comparisons between the rate of a chemical reaction and the rate of a physical change such as evaporation. Both depend on temperature and surface area.

Students

Physics

2.23   Factors affecting rate of evaporation

Biology

E1.A   Rates of reaction in the body;
         metabolism and enzymes

3.1   Photosynthesis

5.7   Temperature control in the body

6.1   Vitamins

E6.A   Vitamins

10.1   Preserving food in a 'fridge

Activities

Chemistry

Catalysing a reaction (p. 34)

Using heat to speed up a reaction (p. 38)

Drawing an energy diagram (**TG** Activity 6)

Catalysts at work (**TG** Activity 7)

| 7.5 | **Explaining rates** |

Topics covered: particle-based explanations of factors affecting rates of reaction

A simple kinetic theory model can be used to give qualitative explanations of many chemical and physical phenomena. These include rates of reaction, Brownian motion, diffusion, osmosis, expansion, solid, liquid and gaseous states of matter, and pressure changes in gases.

Students

Physics

2.10   Solids, liquids and gases; Brownian
         motion; diffusion

2.13   Expansion

2.16   Pressure/temperature relationship for a
         gas

2.17   Pressure/volume relationship for a gas

2.23   Evaporation

Biology

2.3   Diffusion; osmosis

8   Air and water

8.1   Air

Topics covered: gases in air; the importance of air; measuring the percentage of oxygen in air

Air is a mixture of gases. Some of these, such as oxygen, carbon dioxide and nitrogen, are essential for life processes. They are kept in approximate balance by the living organisms on Earth which, together, extract them from the atmosphere and return them to the atmosphere. Human activity can upset the balance. There is concern that a build-up of carbon dioxide in the atmosphere may have a long-term effect on world climate.

It is the water vapour content of the atmosphere which shows greatest variation with time and place. Along with temperature and pressure, it is a major factor in determining weather patterns.

Most of the atmosphere is within 10 kilometres of the Earth's surface. The density and pressure of the air normally decrease from sea level upwards.

Students

Physics
1.1   Measuring volume
1.2   Density of air
1.3   Measuring the density of air
1.11   Air resistance
1.19   Atmospheric pressure
1.20   Measuring atmospheric pressure
2.19   Convection in air
2.24   Water vapour in the air
2.25   Comfortable climates; carbon dioxide build-up in the atmosphere

Biology
E1.A   The biosphere
1.8   Fish and amphibians get oxygen from water
3.1   Carbon dioxide needed for photosynthesis
3.4   Transpiration: adding water vapour to the air
E3   The oxygen cycle
4.4   Oxygen needed for respiration
4.5   Mixture of gases in the lungs
9.2   The carbon cycle
9.3   The nitrogen cycle
9.5   The water cycle
9.7   Air pollution
E9.B   Air pollution and the greenhouse effect
E9.C   Air pollution and the ozone layer

Activities

Chemistry
Living in a greenhouse? (**TG** Activity 9)

Biology
Light, carbon dioxide and photosynthesis (p. 18)
Photosynthesis and oxygen (p. 19)
Breathing and gas exchange (p. 31)
Demonstrating respiration (p. 32)

8.2   Making use of air

Topics covered: separating gases from the air by fractional distillation; uses of oxygen, nitrogen and the noble gases

When liquid air is warmed, the different gases in it boil off at different temperatures. The principle is used to separate the gases.

Fuel-burning engines all need a supply of oxygen from the air. Jet engines also take in huge masses of air and eject it backwards to produce a forward force.

Heat can be carried by moving air in the form of convection currents. Materials which trap air are good insulators because air is a poor conductor.

Air provides most of the raw materials from which animals and plants are made. During photosynthesis, plants change substances from the air into food.

Many animal and plants make use of the physical properties of air. Smells are detected because of diffusion through air. Sounds are heard because compression waves can travel through air. Air is dense enough to support the flight of birds and insects. The pollen of many plants is transported by moving air.

8.3   Air pollution

Topics covered: sources of pollution; types of air pollution; reducing pollution

Air is polluted by the exhaust gases from cars, lorries and fuel-burning power stations. It can also be polluted by the fumes from synthetic materials and by nuclear radiation. One problem with air pollution is that it is quickly spread by convection currents in the Earth's atmosphere. The most effective way of dealing with pollution is to prevent the release of pollutants into the air.

Teachers may wish to include air pollution in a general study of pollution issues. These might also include water pollution, noise and the environmental impact of power stations and supply networks.

8.4   Water and the water cycle

Topics covered: water as a compound; tests for water; the water cycle; water in living things; the need for water; uses of water

Living things are mostly water. It is essential for the processes of life. Plants need water for photosynthesis. It is used to transport food and other substances through them, and its pressure keeps them firm or turgid. Animals need water. Their blood and body fluids are mainly water, and the processes which transfer substances between cells depend on water.

On Earth, water is found in solid, liquid and gaseous forms. It is continually recycled through the atmosphere, and is a major cause of weathering in rocks and soil.

Society has found many industrial and technological uses for water. It is widely used as a means of carrying heat. And, as sea, lake or river, it can be used as a source of energy.

Students

Physics
1.2   The density of water
1.3   Measuring the density of water
1.18 Water pressure
2.8   Wave, tidal and hydroelectric energy
2.11 The freezing and boiling points of water
2.15 The expansion of water
2.20 Water as a storer of heat
2.22 Ice and water; change of volume on freezing
2.23 Evaporation and boiling
2.24 Water vapour in the air
2.25 Changing the climate
3.8   Water waves
3.9   Water waves
4.2   Water as a conductor
4.22 Use of water in power stations
5.6   Water as a reactor coolant

Biology
1.8   Fish and amphibians
2.1   Cells are mainly water
2.3   Diffusion and osmosis in cells
3.1   Plants take in water to make food
3.3   Water under pressure keeps plants firm or turgid
3.4   Transpiration stream carries substances through a plant
3.8   Seeds need water for germination
3.9   Roots grow towards water
5.4   Blood is mainly water
5.7   Evaporation used to control body temperature
5.8   Excretion: part of the water cycle, eliminating excess water from bodies
8.4   Water protects a growing embryo
8.6   Lack of rainfall causes crop failure and food shortages
9.4   Water as a weathering agent
9.5   The water cycle
9.6   Water pollution
9.9   The water supply

Activities

Chemistry
Testing for water (p. 12)
Electrolysing water (p. 30)

Biology
Photosynthesis (pp. 17-20)
Transport tissue in plants (p. 23)
Transpiration (p. 24)
Measuring transpiration (p. 26)

8.5	**Our water supply**

Topics covered: tap water; waterworks; waste water and sewage plants

Reservoirs formed behind dams are one source of water for the mains supply. Dam design has to allow for the very high pressures at the base. Some reservoirs can be used as a source of hydroelectric energy. But all such artificial lakes have a major impact on the environment.

Natural water can contain bacteria which are harmful to humans. It may also contain other pollutants. Careful filtering and cleaning is essential before water is piped to homes. Teachers may wish to draw parallels between the function of a waterworks in cleaning natural water and the function of the kidneys in cleaning blood.

Sewage plants are used to treat some waste water and sewage. But untreated effluent can still be a major source of pollution in rivers and on the coast.

Students

Physics
1.18 Pressure behind a dam
2.8   Lakes as an energy resource
4.23 Environmental problems of hyroelectric power

Biology
5.8   The kidneys
6.5 and 6.6   Digestion: source of water
9.5   The water cycle
9.6   Water pollution
9.9   Water supply: sewage treatment
10.1 Bacteria; the need for clean water

8.6	Soft and hard water
8.7	Making hard water soft

Topics covered: causes of soft and hard water; calcium carbonate; temporary and permanent hardness; advantages and disadvantages of hard water; making hard water soft; comparing hardness

The most common cause of hard water is the calcium carbonate which gets dissolved when rain soaks through limestone or chalk soils and rock. This temporary hardness is removed by boiling, but it causes scale to form on water-heating elements and in hot water pipes.

Hard water is a useful source of calcium for bones and teeth. There is also some evidence to suggest that it reduces the risk of heart disease.

Students

Physics
1.1	Measuring volume
2.19	Simple hot water circuit in a house
2.23	Boiling water
4.5	Heating elements

Biology
4.1	Bones
5.3	Heart disease
6.1	Calcium in food
9.4	Soil

Activities

Chemistry
Distilling cola (p. 11)
Investigating the limestone cycle (p. 26)
Making soap (p. 36)

8.8	Water pollution

Topics covered: causes and effects of water pollution

Pollutants in water can reduce its oxygen content so that river life dies. Poisonous chemicals in rivers can pass along food chains so that birds, humans and other fish-eating animals get poisoned.

Polluted water can find its way, through soil, into plants. Potential pollutants include industrial chemical waste and, even more controversially, radioactive waste.

Teachers may wish to include water pollution in a general study of pollution issues. These might also include air pollution, noise and the environmental impact of power stations and supply networks.

Students

Physics
3.16	Noise
4.22	Power stations
4.23	Environmental problems of power generation and transmission
5.2	Nuclear waste
5.3	Nuclear waste dumps
5.5	Nuclear reprocessing plants
5.6	Nuclear power stations

Biology
1.8	Fish
9.1	Food chains
9.4	Plants take materials from soil
9.6	Water pollution
9.7	Air pollution
9.9	Sewage treatment
E9.A	Pollutants in water can enter food chains

Activities

Chemistry
How acid rain kills (**TG** Activity 8)

Biology
Pollution (p. 54)
A fishy problem (**TG** Activity 6)

9 Acids and bases

9.1	**Acids and alkalis**
9.2	**A closer look at acids**
9.3	**A closer look alkalis**
9.4	**Bases and neutralization**

Topics covered: acids; alkalis; pH scale; properties of acids and alkalis; strong and weak acids and alkalis; bases; neutralization

Acids and alkalis both contain ions. The presence of ions increases the electrical conductivity of a solution. Where different solutions have different ion concentrations, a voltage can be measured between them. The principle is used in a pH meter.

Acids have an important part to play in the body. Acid in the stomach helps digest food. Some enzymes work best in acid or alkaline conditions.

Acids can be neutralized by alkalis. The principle is used in insect sting ointments and indigestion tablets. Some plants grow poorly in acidic or alkaline soils. The pH value can be adjusted by treating soil with suitable chemicals.

Exhaust gases from vehicles and power stations cause acid rain which can damage stonework and destroy forests.

Students

Physics

2.4	Engines produce exhaust gases
4.2	Water as a conductor
4.4	Measuring voltage
4.22	Fuel-burning power stations produce exhaust gases
4.23	Acid rain
5.1	Charges in atoms
5.2	Ions

Biology

E1.A	Effect of acid or alkaline conditions on enzymes
1.7	Stinging arachnids and insects
2.6	DNA, an acid
4.4	Anaerobic respiration produces lactic acid
5.8	Excretion: urea a weak alkali
6.1	Food; vitamins
6.6	Acid in the stomach
6.7	Liver: urea a weak alkali
7.1	Detecting acids on the tongue
9.4	Acid in rain dissolves rock; soils can be acid or alkaline
9.5	Rain in the water cycle
9.6	Water pollution

Activities

Chemistry
Investigating the pH scale (p. 18)
Investigating acids (p. 20)
Neutralizing vinegar with lime (p. 22)
Investigating the limestone cycle (p. 26)
Rainbows (p. 48)
How acid rain kills (**TG** Activity 8)

Biology
How to compare soil samples (p. 52)
The effects of soap and pH on bacteria (p. 63)

9.5	**Making salts (I)**
9.6	**Making salts (II)**

Topics covered: making salts from acids; making insoluble salts; making salts from their elements; making salts in industry

Adding salt to water raises its density, lowers its freezing point and increases its electrical conductivity.

Nitrates exist naturally in the soil, mainly as a result of bacterial action and lightning. Huge amounts of nitrates and sulphates are added to soil as fertilizers. Production of salts often involves the evaporation of water, so that only the salt remains.

Students

Physics

1.1	Measuring mass and volume
1.2	Density
1.3	Measuring density
2.22	Salt lowers the freezing point of water
2.23	Evaporation
4.2	Water as a conductor

Biology

4.1	Skeleton: bone is partially a salt: calcium phosphate
6.1	Food: minerals are salts
6.7	Liver: minerals are salts
9.3	Nitrates and the nitrogen cycle
9.4	Soil; fertilizers
9.6	Pollution from fertilizers

Activities

Chemistry
Reacting metals with acids (p. 21)
Making Epsom salts (p. 23)

10 ══════ Metals

| 10.1 | **Metals and non-metals** |

Topics covered: the properties of metals; differences between metals; metals and non-metals compared

Metals have many properties which distinguish them from non-metals, including higher densities and melting points, greater strength, higher reflectivities, and higher electrical and thermal conductivities. Metals are better absorbers of nuclear radiation than most other materials. Some metals are magnetic.

For healthy growth, plants need traces of some metals. They are also needed, in small quantities, by the body. Some, like lead and mercury, are poisonous and can damage the brain and central nervous system.

Students

Physics
1.2	Density
1.16	The strength of metals when stretched
2.11	Temperature
2.13	Expansion of metals
2.18	Metals as good conductors of heat
2.22	Melting point
3.2	Metals used in mirrors
3.3	Metals as reflectors of electromagnetic waves
3.11	Metals as reflectors of heat radiation
3.13	Vibrations produce sound
4.2	Metals (and graphite) as good conductors of electricity
4.5	Different metals have different resistances
4.13	Magnetic and non-magnetic metals
5.2	Metals as absorbers of nuclear radiation (and X-rays)

Biology
E3	Metals and other minerals in plants
4.1	Bones and teeth contain calcium
5.4	Red blood cells contain iron
6.1	Calcium and iron in food
6.7	Liver: calcium potassium and iron compounds
7.6	The brain can be damaged by lead
9.6	Lead and mercury are poisons which can enter food chains
9.7	Petrol companies are reducing the lead in petrol

Activities

Chemistry
Looking at elements (p. 15)
Burning elements in oxygen (p. 16)
Testing for electrical conductivity (p. 28)

Physics
Stretching a spring (p. 15)

10.2	**Metals and reactivity (I)**
10.3	**Metals and reactivity (II)**
10.4	**The reactivity series**

Topics covered: metal/oxygen reactions; metal/water reactions; metal/acid reactions; displacement reactions; the reactivity series; electric cells

Some metals have higher reactivities than others. Metals of different reactivities can be used in cells and batteries to provide a flow of electrons.

Artificial hip joints and bone-reinforcing pins are made from stainless steel so that they do not react with materials in the surrounding tissue.

Students

Physics
| 4.4 | Batteries and cells |

Biology
| 4.2 | Joints and bones in the human skeleton |

Activities

Chemistry
Reacting metals with oxygen (p. 17)
Reacting metals with acids (p. 21)
Electroplating (p. 31)
Best battery (p. 58)
Measuring the volume of gas given off in a reaction (**TG** Assessment 1)

| 10.5 | Metals in the Earth's crust |
| 10.6 | Extracting metals from their ores |

10.7	Making use of metals
10.8	More about aluminium
10.9	More about iron

Topics covered: metals (and non-metals) in the Earth's crust; scarce metals; metal ores; mining; extracting metals; recycling metals

Silicon is as abundant in the Earth's crust as all the metals put together. Silicon is the basic material used in glass and in microchips.

Some ores are magnetized by the Earth's magnetic field. When scrap metals are recycled, electromagnets can be used to separate iron and steel from the rest.

The body needs small amounts of some metals. Plants, which are eaten as food, extract metals and other minerals from the soil.

Topics covered: the uses of pure metals and alloys; the extraction and production of aluminium and iron; steel as an alloy of iron

Metals, with their excellent mechanical, thermal and electrical properties, are the most important of all industrial and technological resources. The properties of any metal can be modified by mixing other substances with it to form an alloy.

Plants need traces of some metals for healthy growth. They are also needed in small amounts by the human body. The metals are present in some foods.

Students

Physics
- 3.4 Glass as a refracting material
- 3.7 Glass used for lenses
- 4.14 The Earth's magnetic field
- 4.16 Electromagnets can separate ferrous and non-ferrous metals
- 5.11 The semiconductor diode — the simplest silicon chip
- 5.13 Transistors are made from silicon
- 5.15 Microchips as logic gates
- 5.17 Microchips as operational amplifiers

Biology
- 3.1 Plants take in minerals including metals
- E3 Minerals absorbed by plants
- 4.1 Calcium is needed for bones and teeth
- 6.1 The body obtains iron and calcium from food
- 9.4 Soil contains minerals including metals

Students

Physics
- 1.2 Densities of metals
- 1.12 Metals used in rocket and jet engines
- 1.16 Strength of metals
- 2.4 Metals used in petrol and diesel engines
- 2.12 Mercury in thermometers
- 2.14 Using expansion
- 2.18 Metals as conductors of heat
- 3.3 Metals as reflectors of electromagnetic waves
- 3.11 Metals as reflectors of heat radiation
- 4.2 Metals as conductors of electricity
- 4.5 Metals as resistors and heating elements
- 4.9 Copper wires in a mains circuit
- 4.11 Copper wires in house circuits
- 4.13 Magnets and magnetic metals
- 4.16 Metals in electromagnets
- 4.18 Metals in motors
- 4.20 Metals in generators
- 4.21 Metals in transformers
- 4.22 Power stations: steel for boilers and turbines; aluminium for overhead cables
- 4.23 Metals for power cables
- 5.2 Metals as absorbers of nuclear radiation
- 5.6 Nuclear reactor: steel pressure vessel and fuel cans
- 5.14 Copper connecting wires in electronic circuits

Biology
- E3 Metals absorbed by plants
- 4.1 Calcium is needed for bones and teeth
- 4.2 Hip joints — artificial ones can be made from stainless steel
- 6.1 The body obtains iron and calcium from food
- 6.4 Teeth — calcium compounds and mercury and gold fillings

Activities

Chemistry
Elecroplating (p. 31)
Taping it (p. 60)
Investigating the electrolysis of compounds (**TG** Assessment 4)

Biology
Bones (p. 27)
How to compare soil samples (p. 52)

10.10   Corrosion

Topics covered: corrosion; protecting metals against corrosion

Metal structures offer great resistance to compression, tension and bending forces. But this resistance can be weakened considerably by corrosion.

Corrosion is a potential problem in power stations where metal surfaces may be exposed to very high temperatures or steam.

Corrosion is indirectly a cause of pollution, in that rusting vehicles and household equipment are sometimes dumped where they can spoil the environment.

11 Hydrogen and nitrogen

11.1 Hydrogen

Topics covered: making hydrogen; properties and uses of hydrogen

Hydrogen is less dense than other gases. In rockets, it is carried in liquid form for use as a fuel. With oil supplies dwindling, cars may eventually burn hydrogen in their engines rather than petrol. This would reduce pollution.

In the Sun, energy is released when hydrogen is converted into helium during nuclear fusion. Fusion reactors, which produce no radioactive waste, may eventually replace fission reactors in power stations.

Students

Physics
1.2   Density
1.12   Hydrogen for rocket engines
2.4   Engines
5.6   Nuclear power

Biology
3.1   Energy from the Sun
9.7   Air pollution
E9.C   Pollutants from burning fossil fuels

Activities

Chemistry
Reacting metals with acids (p. 21)
The hydrogen revolution (**TG** Activity 11)

11.2 Nitrogen

Topics covered: properties of nitrogen; nitrogen needed for proteins in plants and animals; the nitrogen cycle

About four-fifths of the Earth's atmosphere is nitrogen. The element is needed to make the proteins in plant and animal tissue. Nitrogen is continually recycled through the atmosphere, soil, plants and animals. Lightning, bacteria and fertilizers all have an important part to play in this process.

Fuel-burning vehicles and power stations give out oxides of nitrogen with their exhaust fumes. These pollute the atmosphere and cause acid rain.

Students

Physics
1.17   The atmosphere
2.4   Engines produce exhaust gases
4.6   Lightning
4.22   Fuel-burning power stations
4.23   Acid rain

Biology
3.1   Plant roots take in nitrates from the soil
5.8   Waste nitrogen compounds excreted in urine
6.1   Proteins needed for building body tissues
9.3   The nitrogen cycle
9.4   Nitrogen-fixing bacteria in soil; fertilizers
9.7   Air pollution; acid rain
E9.C   Air pollution; acid rain

Activities

Chemistry
Testing foods for nitrogen (p. 33)
How acid rain kills (**TG** Activity 8)

11.3 Ammonia
11.4 Ammonia and nitric acid in industry

Topics covered: making ammonia; the properties of ammonia and ammonia solution; uses of ammonia; making nitric acid from ammonia

In industry, ammonia is made using the Haber process. The ammonia condenses from a mixture of gases and is collected as a liquid. It is mainly used to make nitric acid and fertilizers.

During the production of nitric acid, oxides of nitrogen are formed. These, along with the acid itself, are a potential source of pollution.

Physics

1.2   Density – ammonia is less dense than air

2.10   Molecules in liquids and gases

2.11   Temperature

2.23   Condensation

4.23   Acid rain

Biology

9.4   Fertilizers – many are made using nitric acid

9.6   Nitric acid is a potential water pollutant

9.7   Oxides of nitrogen cause acid rain

E9.C   Oxides of nitrogen cause acid rain

10.1   Use of ammonia as a disinfectant

10.3   Tobacco smoke contains ammonia

11.5   Fertilizers

Topics covered: population growth, why fertilizers are needed; what plants need from soil; fertilizers; fertilizers as a source of pollution

With intensive farming methods, world food production is enough to meet world needs. But distribution is unequal, and production methods in many areas inadequate.

Plants need a range of elements from the soil. If high crop yields are required, fertilizers must be added to the soil to keep it in good condition. Excessive use of fertilizers causes pollution. Some argue that 'natural' farming methods are preferable, even if yields are lessened as a result.

Students

Physics

2.8   More energy required to manufacture more fertilizers for growing agriculture industry

Biology

3.3   Plant roots need elements from the soil

8.6   Population growth; food distribution problems

E8.B   Over-use of fertilizers

9.4   Fertilizers in soil

9.6   Fertilizers as a cause of water pollution

E9.A   Water pollutants can enter food chains

Activities

Chemistry

Testing foods for nitrogen (p. 33)

Making a fertilizer (p. 37)

Biology

How to compare soil samples (p. 52)

11.6   A look at a fertilizer factory

Topics covered: the fertilizer production process; choosing a site for a factory

Teachers may wish to contrast the issues surrounding the siting of a fertilizer factory with those surrounding the siting of power stations.

Students

Physics

4.22   Power stations

5.6   Nuclear power stations

Biology

9.7   More about pollution

10.9   Conservation and ecology

12 Oxygen, sulphur and chlorine

Oxygen is essential for life. It is needed for respiration in plants and animals. Oxygen balance is maintained in the atmosphere because, overall, animals use up oxygen and plants produce it.

Oxygen is needed for combustion in engines, and in fuel-burning power stations and heating systems. Unless there are regular changes of air in a house, the occupants may be deprived of oxygen and poisonous pollutants may accumulate.

Students

Activities

12.2 | Oxides
12.3 | Sulphur and sulphur dioxide

Oxides form when substances burn in oxygen. Some of the oxides in the exhaust gases from vehicles and power stations cause pollution.

Sulphur dioxide can damage the lungs. It is a major cause of acid rain. The sulphur content of fuels is reduced to lessen the problem.

In small quantities, sulphur dioxide is used as a preservative in food and drink.

Students

Activities

Chemistry
Burning elements in oxygen (p. 16)
Investigating carbon dioxide (p. 24)
How acid rain kills (**TG** Activity 8)

12.4	**Sulphuric acid and sulphates**

Topics covered: the Contact process for making sulphuric acid; uses and properties of sulphuric acid; sulphates

Most sulphuric acid is made by the Contact process. This requires precise measurement and control of temperature. In sulphuric acid plants, it is essential to stop oxides of sulphur escaping as these would cause serious pollution. Sulphuric acid is used in car batteries. Sulphates are used as fertilizers.

Students

Physics
2.11   Temperature
2.12   Thermometers and temperature control
4.2   Some batteries contain sulphuric acid
4.23   Acid rain

Biology
9.4   Fertilizers, some are made using sulphuric acid
9.7   Sulphur dioxide causes acid rain
E9.C   Sulphur dioxide causes acid rain

Activities

Chemistry
Looking at elements (p. 15)
Reacting metals with acids (p. 20)
Making Epsom salts (p. 23)
Making a fertilizer (p. 37)
How acid rain kills (**TG** Activity 8)

12.5	**Chlorine**

Topics covered: how chlorine is made; properties and uses of chlorine

Chlorine is less dense than air, so, in laboratory experiments it can be collected by upward displacement of air.

Chlorine is used to sterilize drinking water, and the water in swimming pools. It is also the active agent in sterilizing fluids for babies' bottles.

Students

Physics
1.2   Density
1.3   Measuring density

Biology
8.6   Population: clean water needs chlorine
E8.A   Disadvantages of bottle feeding
9.6   Pesticides: compounds containing chlorine
10.1   Health; preventing the spread of infection

12.6	**Hydrogen chloride, hydrochloric acid, and chlorides**

Topics covered: hydrogen chloride and its properties; hydrochloric acid and its properties; sodium chloride; test for chlorides

Hydrogen chloride is less dense than air, so, in laboratory experiments it can be collected by upward displacement of air. In the stomach, hydrochloric acid kills germs and helps enzymes to work. Sodium chloride is used to melt ice on the roads.

Sodium chloride improves the flavour of food. It provides the sodium needed for body fluids. But an excess may cause high blood pressure. It has also been linked with certain types of cancer.

Students

Physics
1.2   Density
1.3   Measuring density
1.19   Pressure: effect of HCl dissolving in the fountain experiment
2.22   Salt lowers the freezing point of water

Biology
5.3   Heart disease
5.7   Salt is lost from the body during sweating
6.2   Salt in food may bring health risks
6.3   Salt is a food additive
6.6   Hydrochloric acid in the stomach
E10.B Causes of cancer

Activities

Chemistry
Separating salt from sand (p. 9)
Investigating acids (p. 20)
Reacting metals with acids (p. 21)
Investigating carbon dioxide (p. 24)
Light chemistry (p. 62)

13 Carbon and carbon compounds

13.1 | Carbon and carbonates

Topics covered: carbon; carbon compounds; carbonates and properties

Carbon is found in the proteins and carbohydrates and other compounds which make up living organisms.

Carbon, as graphite, is used in electric motors and generators, and as a moderator in some nuclear power stations. Carbon, as diamond, is the hardest natural material known. Reflection and refraction in diamonds give them their great beauty (and value).

Calcium carbonate is used in making glass. Calcium carbonate, as limestone or marble, has also traditionally been used as a building material. It has great strength under compression, but much less under tension. Modern structures are given tensile strength by adding steel reinforcement.

As a base, calcium carbonate is spread on soils to neutralize any acidity. Calcium hydroxide is used to neutralize liquid acid waste from factories.

Bicarbonate of soda gives off carbon dioxide gas when heated. The expanding gas makes cakes and bread rise in the oven.

Students

Physics
1.16	Strength of materials
2.16	Expansion of gases
3.4	Refraction by glass; refraction produces a spectrum
3.5	Total internal reflections give diamonds their sparkle
4.2	Carbon as a conductor
4.18	Carbon (graphite) brushes in electric motors
4.20	Carbon (graphite) brushes in generators
5.6	Graphite used as a moderator in a nuclear reactor

Biology
3.1	Plants take in carbon dioxide during photosynthesis
6.1	Proteins and carbohydrates in food
9.2	The carbon cycle
9.4	Rock particles in soil; treating soils
9.6	Factory acids as a source of water pollution

Activities

Chemistry
Looking at elements (p. 15)
Investigating the limestone cycle (p. 26)

13.2 | The oxides of carbon

Topics covered: carbon dioxide; properties and uses of carbon dioxide; test for carbon dioxide; carbon monoxide

Carbon dioxide exists as a solid (dry ice) beneath $-78\ °C$. As a gas, it has a greater density than air and will not support burning. Carbon dioxide extinguishers are used for dealing with electrical fires.

The combustion of fuels in vehicles and power stations produces carbon dioxide. Incomplete combustion of fuel in engines produces carbon monoxide. This pollutes the air. It is poisonous because it reacts with haemoglobin.

Students

Physics
1.2	Density
1.3	Measuring density
2.4	Combustion in engines produces carbon dioxide
2.10	Molecules in solids and gases
2.11	Temperature
4.6	Dangers from electrical fires
4.22	Combustion in power stations produces carbon dioxide
5.6	Carbon dioxide as a coolant in nuclear reactors

Biology
4.4	Respiration in humans produces carbon dioxide
4.5	Carbon dioxide breathed out from the lungs
5.4	Haemoglobin in blood cells
9.7	Air pollution
10.3	Carbon monoxide from cigarette smoke

Activities

Chemistry
Investigating carbon dioxide (p. 24)

13.3	Carbon in living things

Topics covered: photosynthesis; respiration; the carbon cycle; the greenhouse effect

Carbon is continuously recycled through living organisms and the atmosphere. Respiration in animals and plants puts carbon dioxide into the atmosphere. Photosynthesis in plants takes carbon dioxide out of the atmosphere. Carbon dioxide is used in building the proteins, carbohydrates and other carbon compounds in all living organisms.

The combustion of fuels in vehicles and power stations is increasing the amount of carbon dioxide in the atmosphere. The slow build-up of carbon dioxide is causing the Earth's atmosphere to warm. This is the greenhouse effect.

Students

Physics

2.4 Combustion in engines produces carbon dioxide

2.8 Energy for plants, animals and fuels

2.25 Carbon dioxide and the greenhouse effect

3.10 Radiant heat is infrared radiation

4.22 Combustion in power stations produces carbon dioxide

Biology

E1.B Carbon dioxide needed for life on Earth

2.3 Waste carbon dioxide removed from cells by diffusion

3.1 Plants take in carbon dioxide during photosynthesis

4.4 Carbon dioxide produced during respiration

4.5 Carbon dioxide breathed out from lungs

E4 Respiration

5.5 Blood carries waste carbon dioxide from cells

E5.A Transport of carbon dioxide from cells

6.1 Food: carbohydrates

9.2 The carbon cycle

E9.B Carbon dioxide and the greenhouse effect

Activities

Chemistry
Fizz facts (p. 53)
Eggs and shells (p. 61)
Living in a greenhouse? (**TG** Activity 9)

Biology
Bones (p. 27)
Light, carbon dioxide and photosynthesis (p. 18)
Demonstrating respiration (p. 32)

13.4	The fossil fuels

Topics covered: coal, oil and gas; producing electricity; pollution from combustion products; other uses of fossil fuels; world reserves of fossil fuels

Fossil fuels store energy which was absorbed from the Sun over 300 million years ago.

In advanced industrial societies, fossil fuels are the principal source of energy. Much of this energy is supplied as electrical energy from power stations.

The use of fossil fuels causes pollution and other environmental problems. The world's supplies of fossil fuels are dwindling.

Students

Physics

1.12 Jet engines burn fuel

2.1 Measuring energy; chemical energy in fuels

2.2 Energy converters

2.4 Petrol and diesel engines burn fuel

2.5 Efficiency of engines as energy converters; power

2.8 The world's energy resources; alternatives to fossil fuels

4.19 The principles of generation

4.20 Generators

4.22 Generating and transmitting power

4.23 Power generation and supply – environmental problems

5.6 Nuclear power

Biology

3.1 Plants absorb solar energy during photosynthesis

8.6 Population growth: increasing need for energy

9.7 Combustion products cause air pollution

E9.B Air pollution; the greenhouse effect

E9.C Air pollution; acid rain

Activities

Chemistry
Comparing fuels (p. 40)
Cracking oil (p. 44)
The energy bill (**TG** Activity 4)
How acid rain kills (**TG** Activity 8)
Living in a greenhouse? (**TG** Activity 9)

13.5 Alternative sources of energy

Topics covered: renewable and non-renewable energy sources; wind, solar and tidal power; hydroelectric power; geothermal energy; biomass

With the world's supplies of fossil fuels dwindling, the search is on for viable alternative sources of energy.

One way of conserving valuable energy resources is to develop methods of being less wasteful with energy. Much can be done to reduce heat losses from the home, but there is an upper limit to the efficiencies of fuel-burning vehicles and power stations.

Students

Physics
- 1.18 Dam construction
- 2.1 Measuring energy
- 2.2 Energy changes
- 2.3 Calculating potential energy (e.g. water in a lake); calculating kinetic energy (e.g. moving water or air)
- 2.5 The efficiencies of engines and power stations; the difference between energy and power
- 2.8 The world's energy resources; alternative sources of energy
- 2.9 Energy saving in the home; problems associated
- 2.20 Water as a carrier of heat energy (e.g. from hot rocks or a solar panel)
- 3.1 Light as energy
- 4.22 Fuel-burning power stations
- 4.23 Environmental problems with hydroelectric schemes
- 5.3 Environmental problems with nuclear power
- 5.6 Nuclear power

Biology
- 3.1 Photosynthesis supplies energy for biomass
- 6.8 Biotechnology: new fuels
- 9.8 Alternative sources of energy
- 9.10 Conservation and ecology: affected by dams and tidal barrages

Activities

Chemistry
Comparing fuels (p. 40)
The hydrogen revolution (**TG** Activity 11)

Biology
An alcohol problem (**TG** Activity 3)
Where there's muck, there's money!
(**TG** Activity 7)

13.6 Alkanes

Topics covered: organic compounds; alkanes; isomers

Car engines (e.g. in taxis) can be modified to burn propane gas rather than petrol. Butane is also a useful portable fuel, but it has too high a boiling point for wider use.

Students

Physics
- 2.1 Chemical energy in fuels
- 2.4 Petrol engines; propane can be used as an alternative fuel
- 2.23 Vapour; boiling point
- 4.2 Alkanes are non-conductors

Biology
- 9.2 Carbon cycle − alkanes form from the decay of dead plants and animals
- 9.9 Sewage treatment works produce methane

13.7 Alkenes

Topics covered: Alkenes; identifying an alkane or alkene; polymerization

Alkene molecules can add on to each other to form polymers. Polythene is the most widely-known polymer. It is easy to mould and is a plastic. It is an insulator and becomes negatively charged when rubbed with a cloth.

Students

Physics
- 1.16 Elastic and plastic materials
- 4.1 Rubbing polythene gives a negative charge
- 4.2 Polythene is an insulator

Biology
- 5.3 Heart disease: vegetable oils contain unsaturated fats similar to alkenes

Activities

Chemistry
Cracking oil (p. 44)

13.8 Alcohols

Topics covered: alcohols; ethanol and methanol; ethanol in drinks

Alcohol can be used as a fuel. In Brazil, car engines have been converted to run on alcohol rather than petrol. The alcohol (ethanol) in alcoholic drinks is made by enzyme action. Enzymes are natural catalysts with a variety of functions. The body uses enzymes to digest food.

Alcohol is a poison. An excess of alcohol can damage the liver and cause heart disease. It has also been linked with certain types of cancer.

Students

Physics
1.1   Units for measuring volume of alcoholic drinks
2.4   Fuel-burning engines
2.8   Biomass energy e.g. alcohol from fermented sugar cane
2.12   Alcohol is used in some thermometers
2.23   Boiling point (alcohol boils at 78 °C)

Biology
4.4   Anaerobic respiration
5.3   Heart disease
6.5   Enzyme action in the digestive system
6.7   The liver — an excess of alcohol causes liver damage
6.8   Biotechnology
10.1   Bacteria
10.4   Alcohol abuse
E10.B   Alcohol and cancer

Activities

Chemistry
Comparing fuels (p. 40)

Biology
An alcohol problem (**TG** Activity 3)

13.9 Chemicals from oil

Topics covered: oil from the Earth; refining oil; cracking hydrocarbons

The energy in oil-based fuels originally came from the Sun. It was absorbed by plants which existed over 300 million years ago. Oil is a mixture of compounds. These all have different boiling points and can be separated by fractional distillation.

Compounds in oil are widely used as fuels, lubricants, road coverings, and as raw materials for producing plastics.

Students

Physics
1.11   Friction is reduced by oil and grease
1.12   Jet engines use kerosene as fuel
2.8   Energy in fuels came from the Sun
2.13   Bitumen as an expansion-gap filler
2.19   Some heating fuels come from oil
2.21   Weatherproof clothing — made from oil
2.23   Evaporation and boiling
4.22   Some power stations burn oil

Biology
3.1   Plants absorb energy from the Sun during photosynthesis
9.6   Pollution caused by crude oil spillages

13.10 Plastics

Topics covered: examples of plastics; advantages and disadvantages of plastics; thermoplastics and thermosets

Plastics are not as strong as metals, but they are cheaper and easy to mould. They can be transparent or opaque. They are widely used for their insulating properties — both thermal and electrical.

Most plastic packaging is non-biodegradable and, as waste, pollutes the environment. The fumes from burning plastics are very dangerous. Some plastics can be recycled.

Students

Physics
1.16   Plastic and elastic materials; strength of materials
2.10   Attractions between molecules in solids
2.18   Plastics are poor conductors of heat
2.21   Weatherproof clothing — made from plastics
2.9   Harmful gases in the home — burning plastics are one source
3.7   Lenses can be made from plastics
4.2   Plastics are insulators
4.6   Dangers from electrical fires
4.9   Plastics in cable insulation and plugs

Biology
5.3   Plastics can be used in replacement-part surgery because they are unreactive
9.7   Pollution — plastic bags source of litter, burning plastics give off harmful gases
9.8   Some plastics can be recycled

ACTIVITY SHEETS

Energy – what is it?

Energy is what makes things light up, get warm, make noise, move. It comes in many different forms, like these.

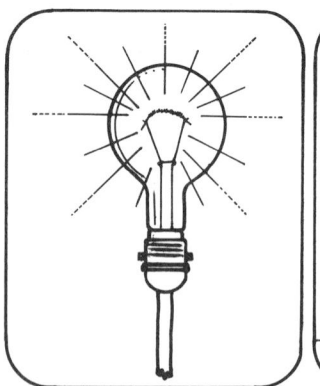

light energy

heat energy

sound energy

kinetic energy – the energy of movement

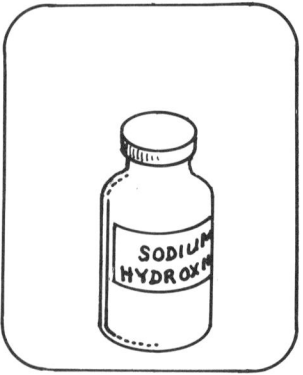

chemical energy – stored in chemicals, fuels, foods

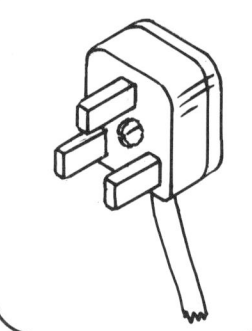

electrical energy

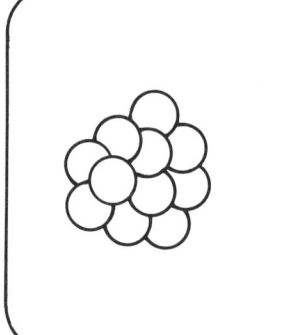

nuclear energy – locked up in the nucleus of an atom

potential energy – due to position, like the energy in a tightly coiled spring

When you eat, chemical energy in the food changes into heat energy which warms you up, and into kinetic energy when you move around. Energy can change from one form to another . . . and that's what science is all about.

Now answer these questions.

▶ Give one example of each energy change:
chemical energy into electrical energy
electrical energy into sound energy
heat energy into sound energy.

▶ A firework has chemical energy stored inside it. What energy changes take place when you light the firework?

▶ Describe how nuclear energy gets changed into electrical energy.

Energy units

You have seen that energy can take different forms, such as heat, light and electricity. It is also measured in different units — or is it?

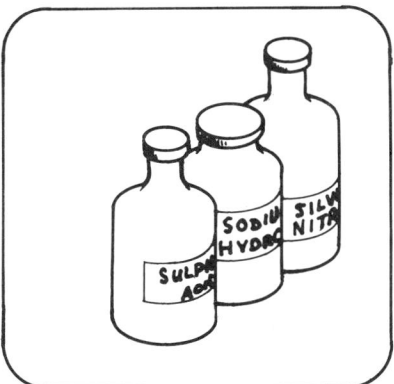
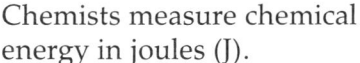

Chemists measure chemical energy in joules (J).

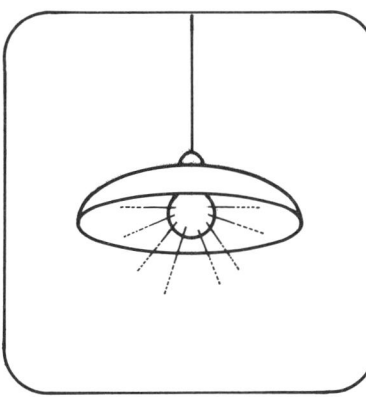

The energy a light bulb uses is measured in watts (W).

Food scientists measure energy in calories (cals).

Confusing? In fact it's not as difficult as it looks. The joule is the standard unit for measuring energy in **any** form. But a joule is only a small amount of energy — about how much you'd use writing two lines in your exercise book. Bigger units are needed.

a thousand joules = 1 kilojoule (kJ)
a million joules = 1 megajoule (MJ)

But what about calories and watts?

1 calorie = 4.2 joules
1 watt = 1 joule per second

Now answer these questions.

▶ Food labels often tell you how many calories the food contains. Why do you think this is?

▶ There are 300 calories of energy in a Mars bar. How many joules is this?

▶ When you eat a Mars bar, what happens to the energy in it?

▶ How many joules does a 60 W bulb use in 1 second?

▶ How long would it take a 60 W bulb to use energy equivalent to that in a Mars bar?

▶ When 1 gram of carbon is burned, it gives out 33 kJ of heat. Does a gram of carbon give more energy or less energy than a Mars bar? Explain.

▶ Why don't we eat carbon to give us energy?

The Sun – the source of it all

We get energy from burning fossil fuels and eating food. And it's all thanks to the Sun! Lets see how much you know about it.

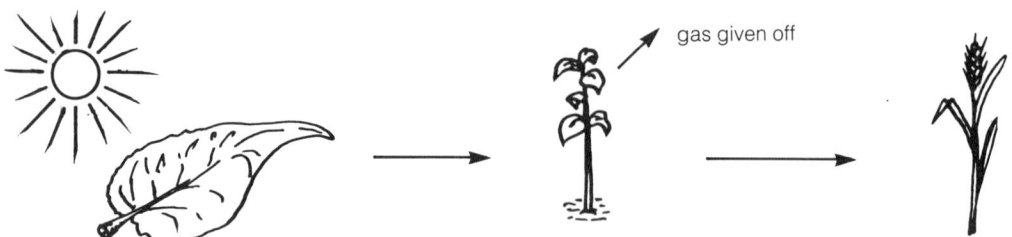

gas given off

▶ Energy from sunlight is trapped in leaves by a green substance called .
_____ .

▶ Plants turn the light energy into chemical energy by making glucose from _____ and _____ . This process is called
_____ .

It gives off a gas called
_____ .

▶ Plants use the glucose to make the things they need: _____ and proteins.

FOOD | OR FOSSILIZATION

We eat plants or animals fed on plants.

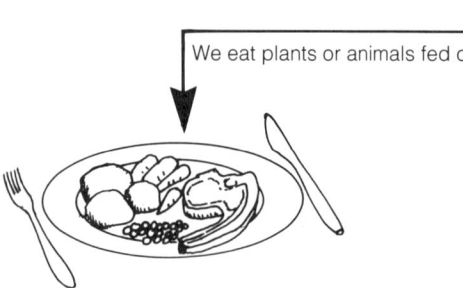

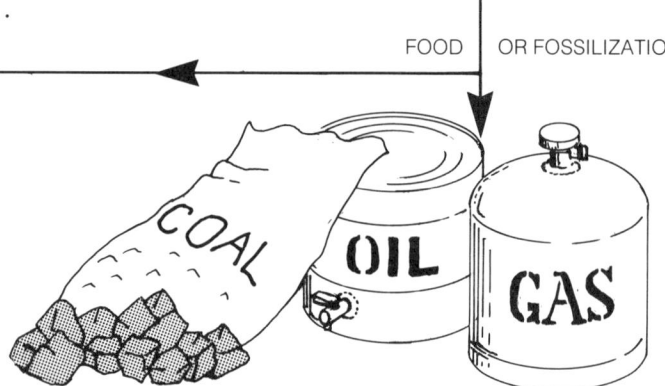

▶ In our bodies, we break carbohydrates back down to glucose again. This process is called _____ .

▶ Over millions of years, the remains of plants, and tiny animals fed on plants, got fossilized. They turned into coal, _____ and _____ , containing the elements _____ and hydrogen.

▶ The glucose goes to our cells, where it reacts with oxygen to form _____ and _____ , This process is called _____ . This is the reverse of _____ . It releases the energy that was trapped from sunlight in the first step. We use this energy for growth, warmth and movement.

▶ Fossil fuels burn in oxygen to form _____ and _____ . This reaction is called _____ . It releases the energy that was trapped from sunlight in the first step. We use this energy for heat, light and movement.

The energy bill

To heat homes we use coal, gas, heating oil, or electricity. But which of them gives most energy for your money? It's up to you to find out.

Source of energy	Units it is sold in	Cost of one unit	How many units for £1?	How much energy for £1?
coal	tonnes			
gas	therms			
heating oil	litres			
electricity	kilowatt hours (kWh)			

▶ Find out the cost per unit of the energy sources listed above. Write your answers in the third column of the table.

▶ Calculate the amount of each you could buy for £1. Write this in the fourth column.

▶ The table shows the energy per unit of each source, in millions of joules. Use it to work out the amounts of energy you get for £1. These go in the fifth column.

How much energy?	
1 kg of coal	30 MJ
1 therm of gas	101 MJ
1 litre of oil	37 MJ
1 kWh of electricity	3.6 MJ

MJ stands for megajoules or 1 million joules

▶ Which source gives *most* energy for £1?

▶ Write down two advantages of this energy source, apart from the cost.

▶ Now write down two of its disadvantages.

▶ Which source gives *least* energy for £1?

▶ Suggest a reason why this source is so expensive.

Energetic chemicals

When you burn a firework, the chemical energy in it is turned into heat, light, sound and movement. But what exactly is this chemical energy? And where is it stored?

You can think of chemical energy as the energy in the bonds holding atoms or ions together. Energy is needed to break these bonds. The same amount of energy is released when they form. This **bond energy** is measured in kilojoules (kJ).

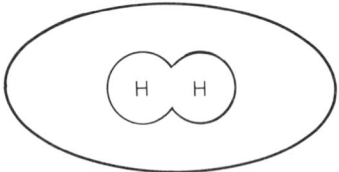 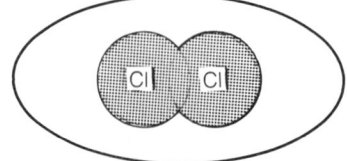 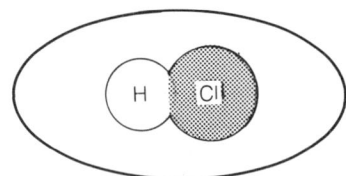

The bond energy of 1 mole of hydrogen molecules is 436 kJ. | The bond energy of 1 mole of chlorine molecules is 248 kJ. | The bond energy of 1 mole of hydrogen chloride molecules is 433 kJ.

Hydrogen burns brightly in chlorine:
$$H_2 (g) + Cl_2 (g) \rightarrow 2HCl(g)$$
Look what happens to the bonds, during the reaction:

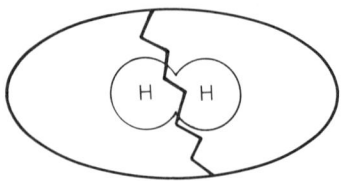 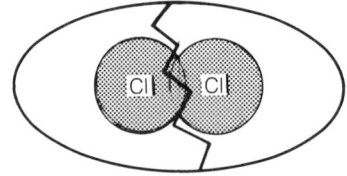 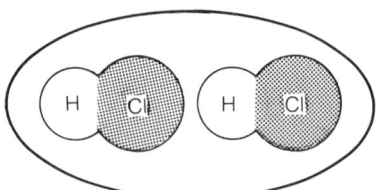

The hydrogen molecules get split into atoms. This *uses* 436 kJ of energy. | The chlorine molecules get split into atoms. This *uses* 248 kJ of energy. | The hydrogen and chlorine atoms bond to form *two* moles of hydrogen chloride. 866 kJ of energy is *released*.

Now suppose a mole of hydrogen gas and a mole of chlorine gas are mixed, and react together. Answer these questions.

► How much energy is needed altogether, to break the bonds in the hydrogen and chlorine?

► How much energy is released when the hydrogen chloride forms?

► Overall, is energy taken in or given out during the reaction? Explain.

► What do you think happens to this energy?

► A mixture of hydrogen and chlorine does not react unless heated. Why is this?

When you burn magnesium ribbon, you get a lot of fizz and sparkle. But what about the energy?

> The reaction between magnesium and oxygen:
> $2\ Mg(s) + O_2(g) \rightarrow 2\ MgO(s)$

Suppose you burn two moles of magnesium (48 g) in a mole of oxygen gas (32 g). This is what happens.

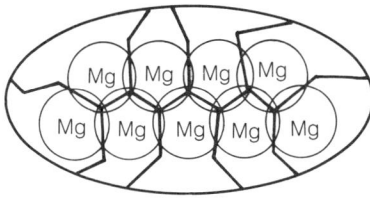 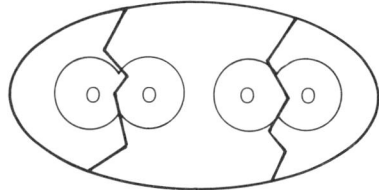 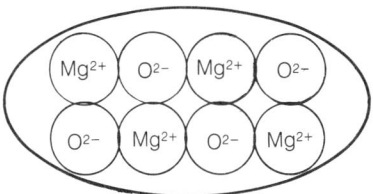

The bonds in magnesium get broken. This *uses* 296 kJ of energy.

The bonds of oxygen get broken. This *uses* 498 kJ of energy.

Magnesium and oxygen bond together. This *releases* 2000 kJ of energy.

The energy change can be shown like this.

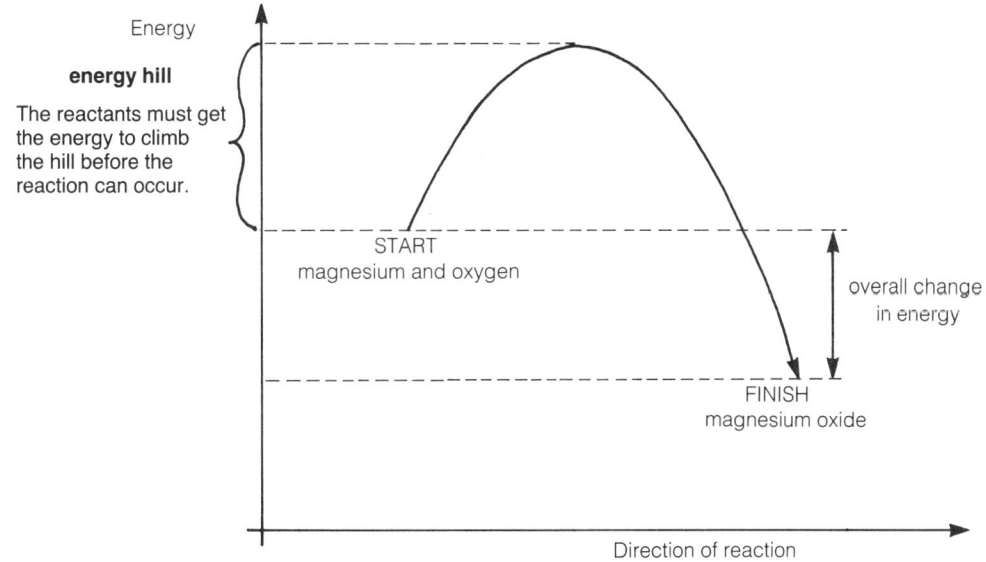

Now try these questions.

▶ In the reaction above:
_____ kJ of energy was used to break bonds.
_____ kJ of energy was released when new bonds formed.

▶ Overall, the reactants must get energy to climb the energy hill *before* the reaction can

start. Where does this energy come from? (**Hint**: how do you start the reaction off?)

▶ Why is energy needed to start the reaction off? (**Hint**: think about the bonds.)

▶ Magnesium burns with a blinding light and a hiss. Where do the light and sound come from?

Catalysts at work

Catalysts speed up reactions without getting used up themselves. How do they work? It's all a question of energy.

This diagram shows the energy changes during a reaction. There is an **energy hill** between the reactants and products. The reactants must get the energy to climb the hill before reaction can take place. We usually provide this energy by heating the reactants up.

Catalysts work by making the energy hill *lower*. So less energy is needed for reaction to occur.

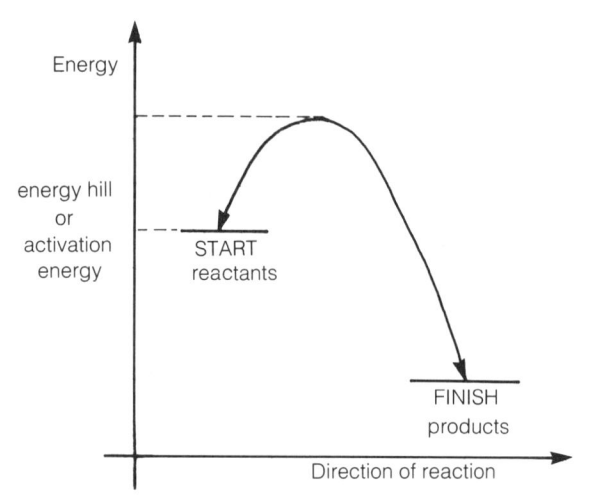

Example 1: A surface catalyst
This drawing shows a catalyst for a reaction between two gases. The gas molecules attach themselves to the surface of the catalyst, and then react with each other.

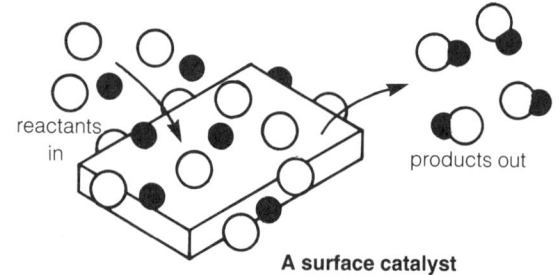

A surface catalyst

Example 2: A biological catalyst
At this very moment, catalysts called **enzymes** are probably at work in your stomach, helping your digestion along. This drawing shows an enzyme in action. Enzymes are destroyed at high temperatures (around 70 °C).

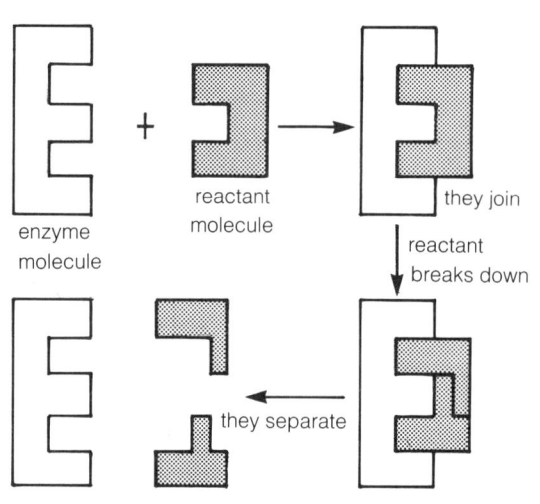

Now try these questions.

► What's the other name for the energy hill?

► Using the idea of an energy hill, explain why heat speeds up a reaction.

► If the energy hill is lowered, but the temperature kept the same, what will happen to the speed of a reaction?

► Try to explain how a surface catalyst speeds up the reaction between gases.

► Try to explain how an enzyme catalyst speeds up a biological reaction.

► Some washing powders contain enzymes. You shouldn't use them in boiling water. Why not?

How acid rain kills

When we burn coal, oil and gas, any sulphur in them forms **sulphur dioxide**. Inside car engines, nitrogen and oxygen combine to form **nitrogen oxides**. Sulphur dioxide and nitrogen oxides are the cause of **acid rain**.

Acid rain has damaged over two-thirds of the conifer trees in Britain. It has killed the fish in over 4000 lakes in Sweden. In Canada over 50 000 lakes are affected. Why is acid rain so harmful?

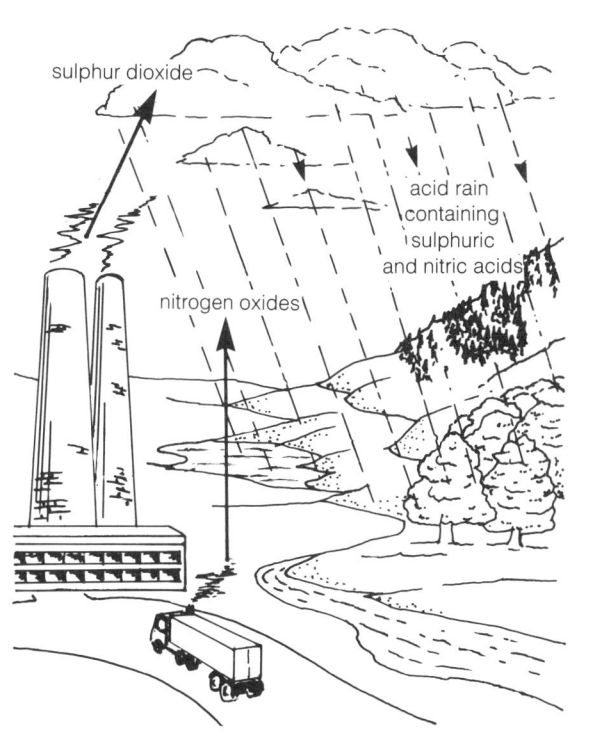

- When acid rain falls on the leaves of a tree, it leaches out the potassium, magnesium and calcium ions the tree needs for food. It also leaches these from the soil round the tree. The soil grows poorer and the tree grows weaker.

- Soil contains aluminium, among other metals. It is present as insoluble compounds which do no harm. But acid rain dissolves these compounds. Aluminium ions are released. They destroy the roots of trees. The trees weaken and die.

- Acid rain soaks through the ground, and into streams and lakes, carrying aluminium ions with it. The acid in the rain can poison fish. But the aluminium is the real killer. It blocks fishes' gills by forming a precipitate of aluminium hydroxide. The fish can't take in oxygen, so they suffocate.

The pH scale of acidity

pH	example
7	pure water
6	'natural' rain-water
5	acid rain has a pH of less than 5.6
4	
3	vinegar
2	lemon juice
1	battery acid

Now answer these questions.

▶ What is the main reason why acid rain kills?

▶ 'Natural' rain is slightly acidic. Why? Would you expect this rain to harm fish?

▶ When acid rain kills trees, birds and insects are also affected. Explain why.

▶ The countries in the table export acid rain to other countries. Draw a diagram to show how this happens.

▶ Write down five things you would do to fight acid rain, if you were a politician.

'Exports' of sulphur dioxide (thousands of tonnes a year)

USSR	7922
UK	3750
E Germany	2888
Poland	2576
W Germany	2338
France	2042
Italy	1804
Belgium	638

Living in a greenhouse?

CHEMISTRY
ACTIVITY **9**

We put carbon dioxide into the air when we burn fuels such as coal and oil. Plants take it from the air and turn it into food, by **photosynthesis**.

The amount of carbon dioxide in the air is tiny. But it is growing steadily. By the year 2000 there will be twice as much as there was in 1850. What will the result be?

Scientists think that the gas will form a blanket round the Earth, letting light in but stopping heat from getting out − just like a greenhouse. The Earth will get warmer. Some think temperatures will rise by 2-3 °C, others by as much as 10 °C. The rise will be greatest around the poles.

A rise of even 2 °C would mean a big change in the weather. There would be less rain in the grain belt of North America, for example. So rain harvests would be smaller. A rise of 10 °C would cause the ice at the poles to melt. Then the oceans would swell, and much of the earth would get flooded.

Year	Level of carbon dioxide
2050	500 ppm?
1988	350 ppm
1958	316 ppm
1900	290 ppm
1850	280 ppm*

*ppm means parts per million

Now answer these questions.

▶ The Industrial Revolution started in Europe around 1800. How did this affect carbon dioxide levels?

▶ More and more of the world's forests are being cut down each year. How does this affect carbon dioxide levels?

▶ In the diagram on the right, the wavy line shows heat rising. What happens to this heat?

▶ In the diagram, which part acts like the glass in a greenhouse?

▶ Suggest one way of reducing the amount of carbon dioxide in the air. How would your method work?

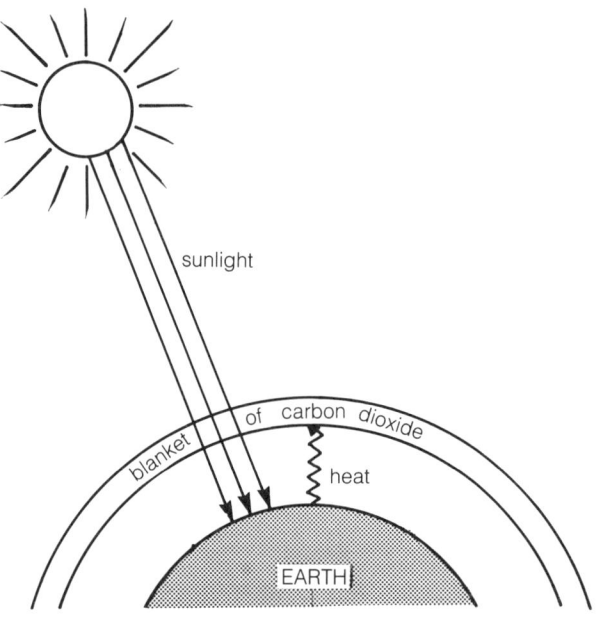

44

About 20 km above the earth, there's a layer of a gas called **ozone**. Ozone is a form of oxygen. But its formula is O_3, not O_2. Where does it come from?

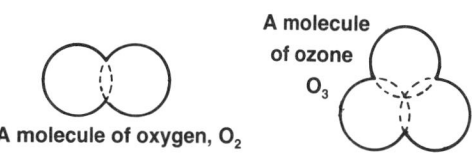

A molecule of oxygen, O_2

A molecule of ozone O_3

High above Earth, energy from the Sun's ultraviolet rays causes oxygen molecules to split into atoms. The atoms collide, and combine again to form ozone. And this protects us. Ozone acts as a filter, soaking up most of the harmful **ultraviolet** (UV) light before it reaches Earth. We would scorch without it.

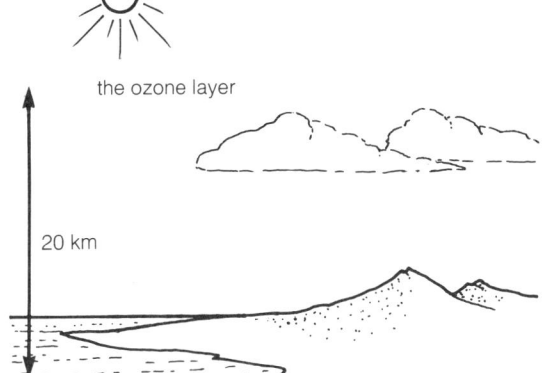

the ozone layer

20 km

But the ozone layer is getting thinner. Large holes are appearing in it. Scientists think this is due to chemicals called **chlorofluorocarbons** (CFCs) which are used in aerosol sprays. CFCs break down in the atmosphere, releasing chlorine. The chlorine then attacks the ozone.

The thinner the ozone layer becomes, the more UV light will reach Earth. Scientists say this will cause a big increase in skin cancers. Crops will be damaged, the Earth will get warmer, and weather patterns will change.

In fact the problem could be so serious that many countries have agreed to cut down on CFCs. In 1988 several UK companies banned their use altogether.

CFCs are used as the carrier gas in aerosol sprays. Check the labels on aerosols to see if they contain CFCs.

Now answer these questions.

► Why is the ozone layer important?

► List things that are sold as aerosol sprays.

► What elements do you think CFCs are made from?

► Suggest five reasons why CFCs might have been chosen for use in aerosol sprays.

► What can *you* do to save the ozone layer?

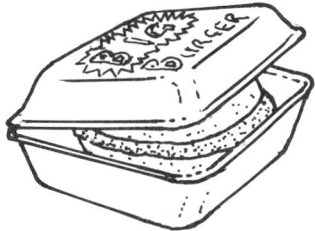

CFCs are also used to make foam cartons for foods like hamburgers and eggs. Why is it bad to burn these cartons?

The hydrogen revolution?

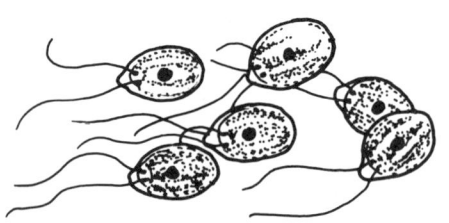

Algae: microscopic water plants that have been used to split water.

Oil supplies are running out. What will we burn in cars and planes when there is no more petrol or jet fuel? One possible answer is **hydrogen**. Unlike oil, there is plenty of hydrogen. But it's locked up in water. How do we get it out?

One way is to use plants. Plants make food by **photosynthesis**, from carbon dioxide and water. In the first stage of this process water molecules are split into hydrogen and oxygen, using energy from sunlight trapped in the plant's cholorophyll.

Scientists have managed to get simple plants called **algae** to split water, without going on to the next stage of photosynthesis. More research is needed to make this method work on a large scale.

Liquid hydrogen produces three times as much energy as the same mass of petrol or jet fuel. It can form an explosive mixture with air. But so can petrol and jet fuel. Scientists think that the safety problem will easily be overcome.

Some alloys of nickel and magnesium will soak up hydrogen like a sponge, then release it when they warm up. Could this be the 'petrol tank' of the future?

Now answer these questions.

▶ Write word equations for:
  (a) splitting water into hydrogen and oxygen
  (b) burning hydrogen in air

▶ Hydrogen can be used over and over again as a fuel. Explain why.

▶ Compared with petrol, hydrogen is a clean fuel. Explain why.

▶ At present, hydrogen is made from North Sea gas. What problem might this cause in the future?

▶ If a plane burns hydrogen instead of petrol, less power is needed for lift-off. Explain why.

▶ To remain liquid, the hydrogen has to be kept under high pressure. Suggest one problem this might cause.

Coffee break

Have you ever wondered how coffee beans end up as instant coffee powder? Read on . . .

Instant coffee was invented in 1938. At that time Brazil was growing more coffee than it could sell. The extra was just thrown away to rot. So the Brazilian government asked food companies to come up with a way of preserving coffee.

The answer – instant coffee. The coffee beans are roasted. Then they're ground into a coarse powder. Water is added and the coffee brewed in giant percolators as high as a three-storey house.

The brew is then treated in one of two ways.

- It can be turned into 'ordinary' instant coffee by spraying it into a heated steel drier. This removes all the water and leaves coffee granules.
- Or it may be turned into 'freeze-dried' instant coffee. For this, the brew is frozen in a giant tank. Then the air is pumped out of the tank. The pressure is reduced until the ice is drawn off as gas, leaving dry coffee granules behind. It is claimed that this freeze-dried coffee has more flavour than the 'ordinary' sort.

Now answer these questions.

▶ In making freeze-dried coffee, ice changes to gas without becoming a liquid. What is this change called?

▶ Why do you think the pressure has to be lowered, to draw ice off as a gas?

▶ To freeze the coffee, the tank is surrounded by liquid nitrogen (boiling point −196 °C). Give three reasons why it's a good choice.

▶ Freeze-dried coffee is more expensive than 'ordinary' coffee. Why do you think this is?

▶ Freeze-drying is used for other foods too. Name three of them. (You can check your answers in the shops.)

Toothpaste to the rescue!

Did you know your mouth is full of bacteria? Many of them live in **plaque**, a moist sticky film which coats your teeth. They feed on the sugar in food, turning it into **acid** which then attacks tooth enamel and dentine. The result is **tooth decay**.

But that's not all. The bacteria in plaque produce poisons which cause gums to swell and bleed. If plaque is not removed, it builds up between your gums and teeth. Your gums get diseased and your teeth may fall out.

Which is where toothpaste comes in. Its job is to get rid of plaque and strengthen tooth enamel. This is what it contains:

decay starting — coating of plaque — enamel — dentine — nerves — gum

A **polishing agent**, finely ground to remove plaque without scratching your teeth. Calcium carbonate is often used as a polishing agent.

A **humectant**. This stops the water in the toothpaste from evaporating when you forget to put the cap on. Glycerol is often used as a humectant.

An **anti-bacterial agent** to slow down the growth of bacteria.

100 ml → **NOPLAK** → The New, Exciting Tasty, Toothpaste

A little **detergent,** to produce foam. This can get into the crevices in teeth and loosen particles of food.

Flavour. Usually spearmint or peppermint, with small amounts of other flavours such as vanilla or menthol added.

A **binding agent**, to hold all the ingredients together in a smooth paste. This can be a natural substance obtained from seaweed, or a man-made product.

A **fluoride compound**. Fluoride helps to strengthen tooth enamel against decay. The compound most often used is sodium monofluorophosphate.

Now try these questions.

► Why is it important to get rid of plaque?

► What does a humectant do? name three other substances that probably contain humectants.

► Is calcium carbonate soluble or insoluble? How does that help it as a polishing agent?

► Calcium carbonate is a base. Why might this help your teeth?

► When you remove plaque, it reforms in 24 hours. How often should you brush your teeth?

► What else may contain fluoride to help teeth?

PRACTICAL ASSESSMENTS

The volume of gas

Measuring the volume of gas given off in a reaction

Teacher/technician notes

Skills

This assessment can be used to test a student's ability in:
a) following instructions *
b) using apparatus and materials *
c) observing safety precautions *
d) measuring accurately
e) recording results
f) handling numerical data
g) interpreting data and drawing an appropriate conclusion.

Abilities marked * have to be checked during the assessment. All other abilities can be checked later from written evidence.

A choice can be made of which abilities are to be assessed.

Previous lessons

Students should be familiar with:
a) the use of cue sheets
b) the reaction of magnesium ribbon with acid
c) assembling apparatus to produce and collect gases
d) the possible use of a measuring cylinder to measure volumes
e) graphical work including the drawing of best fit straight lines
f) the mathematics of proportionality if an advanced conclusion is required.

Apparatus and materials

Each student should have:
  a pair of safety glasses
  three pieces of cleaned magnesium ribbon of length 2.5 cm, 3.5 cm and 4.5 cm.
  a trough (a rectangular plastic box is suitable)
  a measuring cylinder, 50 cm³
  a measuring cylinder, 10 cm³
  a stand, boss and clamp
  a delivery tube, fitted with a bung and with small bends at each end
  a boiling tube
  a boiling tube rack
  a small test-tube (ignition tube)
  30 cm³ of dilute hydrochloric acid, 2 mol dm⁻³
  a piece of graph paper
  a ruler and pencil

At the front of the room:
  trays for dirty apparatus
  trays for completed work

  cue sheets
  spare graph paper
  spare magnesium ribbon and acid
  spare pencils and rulers

Procedure

The following procedure is suggested. Note that most exam boards advise that no more than two or three skills should be assessed in one session.

Time is saved if the equipment is either set out for each student or arranged to be collected at fixed collection points. A list of equipment to be collected may be written on the blackboard.

Following instructions may be made easier by drawing the diagram of the assembled apparatus on the blackboard.

There are three critical stages at which a student could be observed:
a) The student's apparatus should be checked during step 7.
b) The collection of gas should be checked during step 9 to see if the delivery tube is under the boiling tube and the gas is entering the tube.
c) The student should be asked to read the scale on the measuring cylinder when the collection of gas is complete.

Marks should be deducted for breaches of safety precautions and spillage of acid.

Note use of cue sheets and deduct marks where necessary.

Criteria

Following instructions and/or manipulation
There is no need to assess **all** of the following. A choice may be made in deciding whether to assess the *following of instructions*, or *manipulation*, or *both*.
a) safety glasses used
b) correct apparatus used
c) measuring cylinder inverted with exclusion of air
d) magnesium not reacted with acid before the bung is sealed and/or delivery tube is correctly positioned
e) gas collected in measuring cylinder, does not escape outside
f) apparatus manipulated without serious mishaps or spillage
g) correct pieces of magnesium used as revealed by appropriate, and accurate values, for the volumes of gas collected

Observing safety precautions
a) safety glasses correctly worn
b) no spillage of acid, or acid mopped up immediately if split
c) apparatus safely manipulated

Measuring accurately
Each student should be asked to state one or two of the final volumes of gases being collected when asked.

Recording accurately and clearly the results of experiments
The following may be used as a numerical scale. To obtain a suitable scale some of the statements may be either deleted or rearranged.
a) result table with suitable headings i.e. length of magnesium and volume of gas collected, and results entered in increasing value
b) result table with suitable headings and results entered but not in increasing value
c) result table with suitable headings and some results entered
d) result table with suitable headings

e) result table with no headings but completed results in the correct order
f) result table with no headings but completed results in the incorrect order

Handling (or transposing) numerical data
In this experiment a graph is being produced from numerical experimental results.
There is no need to assess **all** of the following.
a) use appropriate scales
b) label axes correctly
c) graph *not* extrapolated through zero
d) correct plotting of two or three points
e) points plotted as crosses or points in circles
f) a straight line drawn
g) a 'best fit' straight line drawn
h) graph drawn in pencil

Drawing appropriate conclusions
Conclude that there is a direct relationship between the volume of gas collected and the amount of magnesium used.

The volume of gas

Measuring the volume of gas given off in a reaction

BACKGROUND INFORMATION

The reaction is between magnesium and dilute hydrochloric acid.

magnesium + hydrochloric acid → magnesium chloride + hydrogen

An excess of acid is used so that all the magnesium reacts. The hydrogen is collected over water in a measuring cylinder. A fixed volume of acid is used but the length of magnesium ribbon used is changed.

Student's instructions

Wear safety glasses during the experiment.

If you have difficulty ask for a cue sheet. A cue sheet is available where you see a star *. There are four sheets available. If you use any cue sheets hand them in at the end of the experiment.

* 1 Fill a trough with water to a depth which will cover a measuring cylinder (50 cm³) placed in the water.
  2 Fill the measuring cylinder with water so that no air remains in it.
  3 Turn the measuring cylinder upside-down in the water. Do not let air in.
  4 Clamp the measuring cylinder with the open end just below the surface of the water. A delivery tube should easily fit beneath it.
  5 Add 10 cm³ of dilute hydrochloric acid to the boiling tube.
  6 Place the 2.5 cm length of magnesium in a small test-tube and place it in the boiling tube of acid. Make sure the acid does not flow into the small tube.
* 7 Connect a delivery tube to the boiling tube. Whilst holding the boiling tube, place the end of the delivery tube under the measuring cylinder in the trough.

  Your teacher must check your apparatus before you do anything else.

  8 Tilt the boiling tube so that the acid runs into the small tube.
  9 Collect and measure the volume of gas produced. Note that the scale of the measuring cylinder is upside-down.
*10 Draw a results table and record the volume of gas collected.
 11 Wash out the boiling tube and the small test-tube.
 12 Repeat the experiment with 3.5 cm and 4.5 cm lengths of magnesium ribbon and fresh acid.
*13 Plot a graph with the **volume of gas** on the vertical axis and the **length of magnesium ribbon** on the horizontal axis.
 14 Use your graph to write a conclusion to the experiment.
 15 Place your results table, graph and conclusion in the tray provided, together with any cue sheets used.
 16 Clear away your apparatus in the trays provided. Clean and tidy your work area.

The volume of gas

Cue Sheet 1:   step 1

Place the measuring cylinder completely under the water. Make sure there is no air in the cylinder.

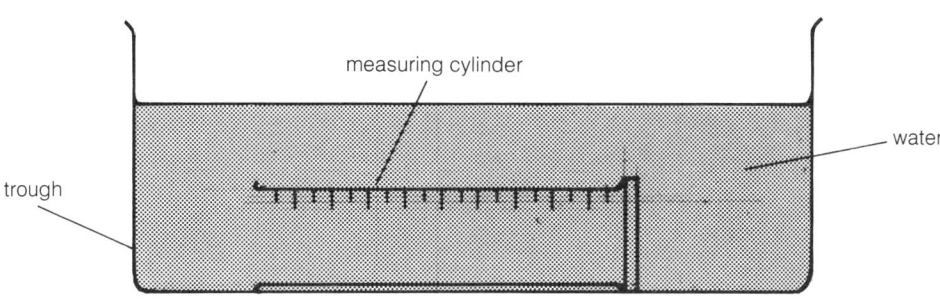

The volume of gas

Cue Sheet 2:   step 7

Tilt the boiling tube upwards so that the acid pours onto the magnesium ribbon.

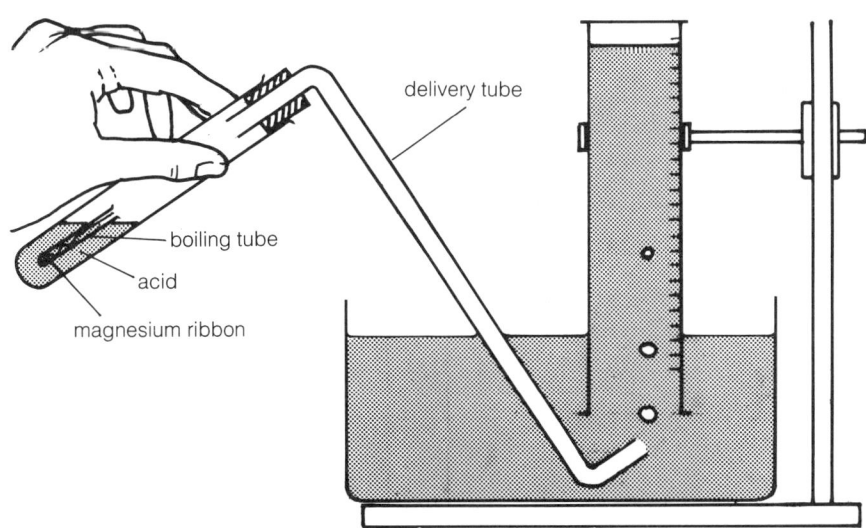

Cue Sheet 3: step 10

Length of magnesium ribbon in cm	Volume of gas collected in cm³
2.5	
3.5	
4.5	

Cue Sheet 4: step 13

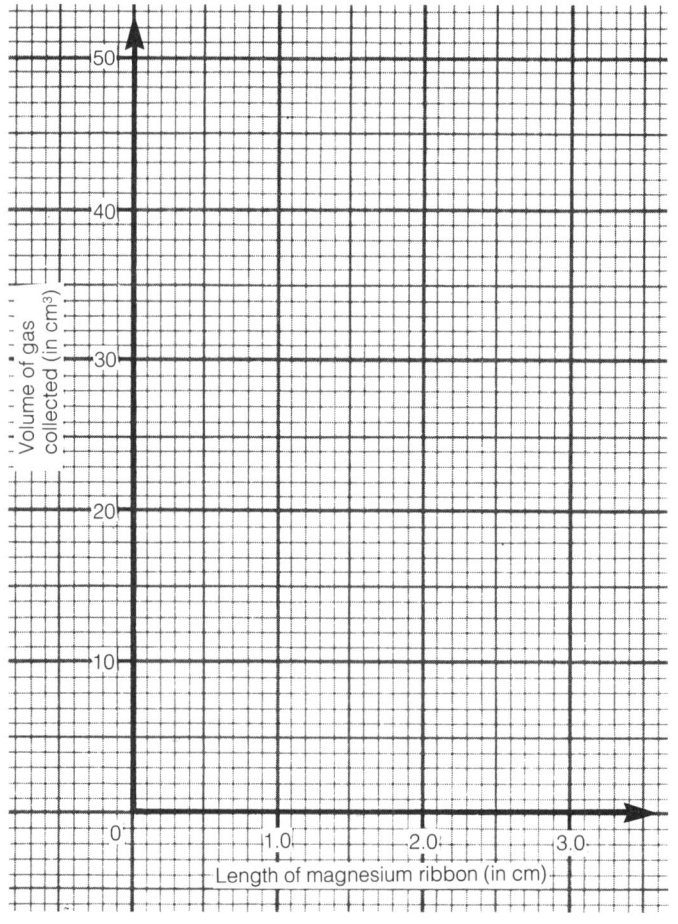

The volume of acid

Measuring the volume of acid needed to neutralize bicarbonate of soda

Teacher/technician notes

Skills

This assessment can be used to test a student's ability in:
a) following instructions *
b) using apparatus and materials *
c) observing safety precautions *
d) observing accurately
e) recording results

Abilities marked * have to be checked during the assessment. All other abilities can be checked later.

A choice can be made of abilities to be assessed.

Previous lessons

Students should be familiar with:
a) the use of cue sheets
b) reaction of sodium hydrogencarbonate with acid
c) acid-base titrations and the use of a burette
d) colours of methyl orange indicator in acid and alkali
e) recording titration results.

Apparatus and materials

Each student should have:
  a pair of safety glasses
  a measuring cylinder, 25 cm^3
  a burette, 50 cm^3
  a burette stand, or stand with boss and clamp
  a burette funnel
  a conical flask, 100 cm^3
  a white tile or white paper
  distilled water
  sodium hydrogencarbonate, 0.2 g samples labelled
    bicarbonate of soda
  dilute hydrochloric acid, 0.1 mol dm^{-3}
  methyl orange indicator solution (screened methyl
    orange may be used)

At the front of the room:
  trays for dirty apparatus
  trays for completed work
  cue sheets
  spare 0.2 g samples of bicarbonate of soda and
    dilute hydrochloric acid

Procedure

Time is saved if the equipment is either set out for each student or arranged to be collected at fixed collection points. A list of equipment to be collected may be written on the blackboard.

Following instructions may be made easier by drawing the diagram of the assembled apparatus on the blackboard.

There are two critical stages at which a student could be observed.
a) The student's burette should be checked after step 4 of the student's instruction.

b) The student should be asked to read the burette at some stage of the titration.

Marks should be deducted for breaches of safety precautions and spillage of acid or bicarbonate of soda.

Note use of cue sheets and deduct marks where necessary.

Criteria

Following instructions
There is no need to assess **all** of the following.
a) safety glasses used
b) correct apparatus used
c) burette clamped securely in an appropriate place
d) air space below the burette tap filled with acid
e) flask swirled during the addition of acid

The following points might be included if emphasis has been placed on them during teaching.
(i)   funnel removed from the burette before the titration commenced
(ii) non-dripping burette used
(iii) the burette tap positioned just inside the flask

Using equipment
The flask handed in may be analysed to see if too much acid has been added by adding drops of sodium hydrogen carbonate solution until the contents are neutral.

The number of drops is an indication of the care and accuracy with which the titration was done. The fewer drops added, the more accurate the titration and the better the manipulation of the student.
A suggested range is:
a) 0 to 3 drops   d) 9 to 12 drops
b) 3 to 6 drops   e) 12 to 20 drops
c) 6 to 9 drops

Observing accurately
One or two burette readings volumes correct.

Recording accurately and clearly the results of experiments
The following may be used as a numerical scale. Some statements may be either deleted or rearranged.
a) result table with suitable headings and volume of acid correctly recorded; one rough titration value and one accurate value correctly entered
b) result table with suitable headings and two titration values correctly entered
c) result table with suitable headings and one titration value correctly entered
d) result table with suitable headings
e) result table with no suitable headings and two titration values correctly entered
f) result table with no suitable headings and one titration value correctly entered

The volume of acid

Measuring the volume of acid needed to neutralize bicarbonate of soda

BACKGROUND INFORMATION

Titration is used to find the concentration of a solution. Antacid tablets contain sodium hydrogencarbonate which is known as bicarbonate of soda. The tablets react with excess dilute hydrochloric acid in the stomach and neutralize it.

sodium hydrogencarbonate + **hydrochloric acid** → **sodium chloride** + **water** + **carbon dioxide**

Student's instructions

Wear safety glasses during the experiment.

If you have difficulty ask for a cue sheet. A cue sheet is available where you see a star *. There are four sheets available. Hand in any cue sheets at the end of the experiment.

  1 Dissolve the bicarbonate of soda (a 0.2 g sample) in about 20 cm³ of distilled water.
  2 Add four drops of methyl orange indicator solution to the bicarbonate solution.
  3 Wash out a burette with distilled water and the wash it out with a small volume (a few cm³) of dilute hydrochloric acid.
* 4 Allow some of the acid to drain through the burette so that the part of the burette below the tap is filled. Fill the burette up to the zero mark or any other number near to zero, such as 1.0, 2.0 or 3.0.
  5 Clamp the burette in a stand.
* 6 Draw a results table. Record the level of acid (in cm³) in the burette.
* 7 Run acid into the conical flask, 2 cm³ at a time. Swirl the flask to mix the solutions. Record the burette reading when the colour in the flask changes.
  8 Wash out the conical flask. Refill the burette with acid.
* 9 Repeat the titration with a fresh supply of bicarbonate solution (20 cm³). This time do it more accurately and find the volume of acid needed to the nearest 0.1 cm³.
 10 Label the flask with your name. Hand it in for an assessment of your titration accuracy.
 11 Place your results table and flask in the trays provided together with any cue sheets used.
 12 Clear away your apparatus in the trays provided. Clean and tidy your work area.

Cue Sheet 1:   step 4

Make sure the space below the tap is filled with liquid.

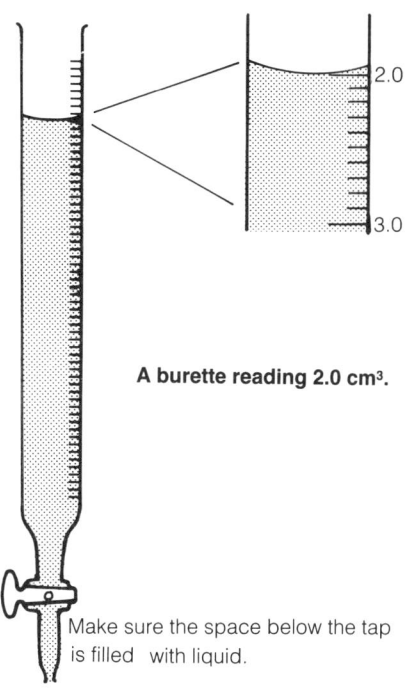

2.0

3.0

A burette reading 2.0 cm³.

Make sure the space below the tap
is filled with liquid.

Cue Sheet 2:   step 6

First reading of burette in cm³	Second reading of burette in cm³	Volume added from burette in cm³
2.0	18.3	16.3

An example is shown in the table. It is *not* a result. Note that readings should be taken correct to one place of decimals.

Cue Sheet 3: step 7

Acid	methyl orange is **red** in acid
Alkali	methyl orange is **yellow** in alkali

The titration is finished when the colour in the flask *just* changes from yellow to red.

bicarbonate (alkali)
+ methyl orange

acid →

yellow   red

Cue Sheet 4: step 9

For an *accurate* titration the volume of acid added must be recorded to the nearest 0.1 cm^3.

For example:
    first reading on burette = 1.0 cm^3
    second reading on burette = 19.8 cm^3
    volume of acid added = 18.8 cm^3

Add the acid 0.1 cm^3 at a time when you are near the end of the titration.

Investigating precipitates

Teacher/technician notes

Skills

This assessment can be used to test a student's ability in:
a) following instructions *
b) using apparatus and materials *
c) observing safety precautions *
d) making accurate observations
e) recording results

Abilities marked * have to be checked during the assessment. All other abilities can be checked later from written evidence. A choice can be made of which abilities are to be assessed.

Previous lessons

Students should be familiar with:
a) the use of cue sheets
b) the meaning of precipitation
c) the poisonous nature of lead compounds
d) filling in a 'league table' of results.

Apparatus and materials

Each student should have:
   a pair of safety glasses
   five test-tubes
   a test-tube rack
   a glass stirring rod
   a test-tube brush
   40 cm^3 of each of the following solutions of:
      copper sulphate, 2 mol^{-3}
      lead nitrate, 1 mol dm^{-3}
      potassium iodide, 1 mol dm^{-3}
      silver nitrate, 0.1 mol dm^{-3}
      sodium carbonate, 1 mol dm^{-3}

At the front of the room:
   trays for dirty apparatus
   trays for completed work
   spare quantities of solutions
   cue sheets

Procedure

The following procedure is suggested. Note that most exam boards advise that no more than two or three skills should be assessed in one session. Time is saved if the equipment is either set out for each student or arranged to be collected at fixed collection points. A list of equipment to be collected may be written on the blackboard.

Following instructions may be made easier by drawing the diagram of the assembled apparatus on the blackboard.

There are no critical stages at which a student should be observed.

Marks should be deducted for breaches of safety precautions and spillage of chemicals.

Note use of cue sheets and deduct marks.

Criteria

Following instructions and/or manipulation
There is no need to assess **all** of the following. A choice may be made in deciding whether to assess the *following of instructions*, or *manipulation*, or *both*.
a) safety glasses used
b) correct apparatus used
c) five test-tubes placed in a test-tube rack
d) five test-tubes one-third full of solution
e) use of glass rod to stir mixtures in the test-tubes
f) apparatus manipulated without serious mishaps or spillage

Observing safety precautions
a) safety glasses correctly worn
b) no spillage of chemicals or mopped up immediately if spilt
c) apparatus safely manipulated

Observing accurately
Assessment can be based on the number of correct observations made.

There are six negative results and 14 positive results. Together with a description of colours this gives a total of 34 observations. Note that 17 observations are being repeated to give the total of 34.

Recording accurately and clearly the results of experiments
The following may be used as a numerical scale. To obtain a suitable scale some of the statements may be either deleted or rearranged. An incomplete table may mean a student has not completed all the parts of the experiment and need not be penalized in this section.
a) result table with the correct additional heading as shown in the student's instructions and correctly filled in using X for a negative result and giving a colour for a positive result
b) result table with the correct additional heading as shown in the student's instructions and incorrectly filled in by not using X for a negative result but giving a colour for a positive result
c) result table with the correct additional heading as shown in the student's instructions and correctly filled in using X for a negative result but not giving a colour for a positive result
d) result table with the correct additional heading as shown in the student's instructions but incorrectly filled in by not using X for a negative result and also not giving a colour for a positive result

Investigating precipitates

BACKGROUND INFORMATION

A precipitate is a solid produced during a chemical reaction in a liquid. It is produced when two solutions are mixed and an insoluble salt is formed.

For example when solutions of sodium chloride and silver nitrate are mixed a white precipitate of silver chloride is formed.

sodium chloride + **silver nitrate** → **sodium nitrate** + **silver chloride**

Student's instructions

Wear safety glasses during the experiment.

If you have difficulty ask for a cue sheet. A cue sheet is available where you see a star *. There are three sheets available. Hand in any cue sheets at the end of the experiment.

* 1 Complete a results table, part of which is shown below.

copper sulphate (aq)				
lead nitrate (aq)				
potassium iodide (aq)				
silver nitrate (aq)				
sodium carbonate (aq)				

2 Wash out five test tubes with tap water and then distilled water. Place them in a test-tube rack.
3 Fill the test-tubes one-third full with copper sulphate solution.
4 Add lead nitrate solution to the copper sulphate solution until the test-tube is two-thirds full. **Take care**. Lead compounds are poisonous. Wash your hands at the end of the experiment.
5 Mix the two solutions by stirring with a glass rod. Remove the glass rod. Wash it with tap water and then distilled water.
6 If a precipitate is formed write the colour in your results table.
   If a precipitate is not formed, write a cross (X) in your results table.
* 7 Repeat steps 1-6 with potassium iodide, silver nitrate, sodium carbonate and potassium iodide solutions.
8 Wash out your five test-tubes. Use a test-tube brush to remove any precipitate which sticks in the test tube.
* 9 Repeat steps 1-8 for all the other possible combinations of solutions shown in the results table.
10 Place your results table, which contains your observations, together with any cue sheets used, in the trays provided.
11 Clear away your apparatus in the trays provided. Clean and tidy your work area.

Investigating precipitates

Cue Sheet 1: step 1

	copper sulphate (aq)	lead nitrate (aq)	potassium iodide (aq)	silver nitrate (aq)	sodium carbonate (aq)
copper sulphate (aq)					
lead nitrate (aq)					
potassium iodide (aq)					
silver nitrate (aq)					
sodium carbonate (aq)					

Investigating precipitates

Cue Sheet 2: step 7

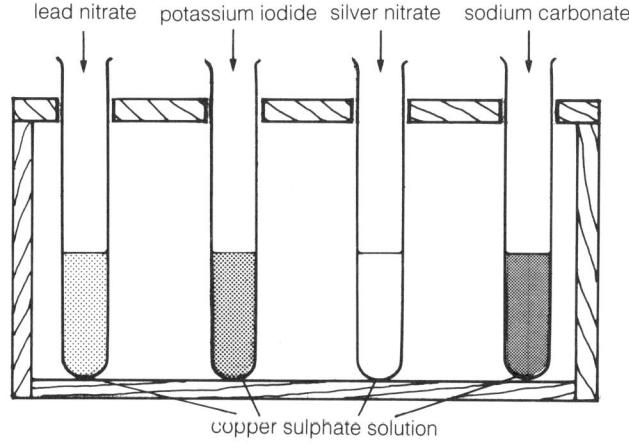

lead nitrate   potassium iodide   silver nitrate   sodium carbonate

copper sulphate solution

Investigating precipitates

Cue Sheet 3: step 9

	copper sulphate (aq)	lead nitrate (aq)	potassium iodide (aq)	silver nitrate (aq)	sodium carbonate (aq)
copper sulphate (aq)		1	2	3	4
lead nitrate (aq)	1		5	6	7
potassium iodide (aq)	2	5		8	9
silver nitrate (aq)	3	6	8		10
sodium carbonate (aq)	4	7	9	10	

Combinations 1-4 have been done.   There are a further 5-10 combinations.
Note that these 10 combinations are repeated in the other half of the table.

Investigating electrolysis of compounds

Teacher/technician notes

Skills

This assessment can be used to test a student's ability in:
a) following instructions *
b) using apparatus and materials *
c) observing safety precautions *
d) making accurate observations
e) recording results

Abilities marked * have to be checked during the assessment. All other abilities can be checked later from written evidence. A choice can be made of which abilities are to be assessed.

Previous lessons

Students should be familiar with:
a) the use of cue sheets
b) constructing simple electrical circuits and knowing how to test a light bulb
c) describing the brightness of a light bulb e.g. as 'bright', 'dim' or 'out'
d) the use of a lab pack if one is being used instead of 6 V battery.

Apparatus and materials

Each student should have:
  a pair of safety glasses
  a lab pack or battery to supply 6 V
  three connecting wires fitted with crocodile clips
  a light bulb in a holder
  two carbon rods
  a beaker, 100 cm³
  solutions of:
    hydrochloric acid, 0.5 mol dm⁻³
    calcium hydroxide, saturated solution
    sodium bromide, 0.5 mol dm⁻³
    potassium iodide, 0.1 mol dm⁻³
    copper sulphate, 0.5 mol dm⁻³

At the front of the room:
  trays for dirty apparatus
  trays for completed work
  cue sheets
  further quantities of solutions
  spare carbon rods, electrical leads and light bulbs

Procedure

The following procedure is suggested. Note that most exam boards advise that no more than two or three skills should be assessed in one session.

Time is saved if the equipment is either set out for each student or arranged to be collected at fixed collection points. A list of equipment to be collected may be written on the blackboard.

Following instructions may be made easier by drawing the diagram of the assembled apparatus on the blackboard.

The student should be observed during step 3, when the electrical circuit should be checked to see that it is correctly connected and the light bulb is working.

Marks should be deducted for breaches of safety precautions and spillage of chemicals.

Note use of cue sheets and deduct marks.

Criteria

Following instructions and/or manipulation
There is no need to assess **all** of the following. A choice may be made in deciding whether to assess the *following of instructions*, or *manipulation*, or *both*.
a) safety glasses used
b) correct apparatus used
c) electrical circuit correctly connected
d) beaker half-filled with solution
e) two carbon rods fully dipped in the solution
f) apparatus manipulated without serious mishaps or spillage

Observing safety precautions

a) safety glasses correctly worn
b) no spillage of chemicals or mopped up immediately if spilt
c) no short circuits connected
d) apparatus safely manipulated

Observing accurately
Assessment can be based on the number of correct observations made.

Recording accurately and clearly the results of experiments
The following may be used as a numerical scale. To obtain a suitable scale some of the statements may be either deleted or rearranged. An incomplete table may mean a student has not completed all the parts of the experiment and need not be penalized in this section.
a) result table with the correct headings i.e. 'bulb', anode and cathode, and all three columns correctly filled in.
b) result table with the correct headings but the anode and cathode columns interchanged, yet all three columns correctly filled in
c) result table with the correct headings but only two columns correctly filled in
d) result table with the correct headings but the anode and cathode columns interchanged and only two columns correctly filled in
e) result table with the correct headings but only one column correctly filled in
f) result table with the correct headings but the anode and cathode columns interchanged and only one column correctly filled in

Investigating electrolysis of compounds

BACKGROUND INFORMATION

When concentrated solutions of salts are electrolysed a metal or hydrogen is produced at the cathode, and oxygen or a halogen set free at the anode. If the metal reacts with water hydrogen is obtained. This does not mean that the metal is actually set free and then reacts. Hydrogen is set free and not the metal.

Student's instructions

Wear safety glasses during the experiment.

If you have difficulty ask for a cue sheet. A cue sheet is available where you see a star *. There are three sheets available. Hand in any cue sheets at the end of the experiment.

1 Collect a 6 V light bulb in a holder and connect it in series with a 6 V battery or power pack set at 6 V. Check that the bulb works.
2 Connect the light bulb in series with a 6 V supply and two electrical leads which have crocodile clips at the ends.
* 3 Connect a carbon rod to each crocodile clip. Put the two carbon rods together to check that the bulb lights.

When you have done this ask your teacher to check your circuit.
Do not proceed further until this has been done.

4 Half-fill a 100 cm³ beaker with dilute sulphuric acid.
* 5 Dip the two carbon rods fully into the acid. Do not let the rods touch each other.
6 Observe what happens around the anode and the cathode. The anode is attached to the + terminal (often coloured red) on the battery.
The cathode is attached to the − terminal (often coloured black) on the battery.
7 Observe the brightness of the light bulb.
* 8 Draw a results table for your observations.
9 Wash out the beaker. Carry out steps 4-7 for the following solutions.
  dilute hydrochloric acid
  calcium hydroxide sodium bromide
  potassium iodide
  copper sulphate
10 Place your results table, which contains your observations, together with any cue sheets, in the trays provided.
11 Clear away your apparatus in the trays provided. Clean and tidy your work area.

Investigating electrolysis of compounds

Cue Sheet 1:   step 3

Place the two rods together. If the bulb does not light, check every connection in the circuit. Replace the crocodile clips if they are rusty.

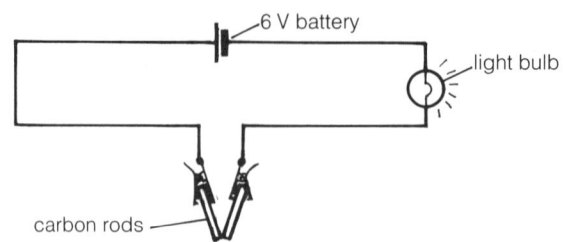

Investigating electrolysis of compounds

Cue Sheet 2:   step 5

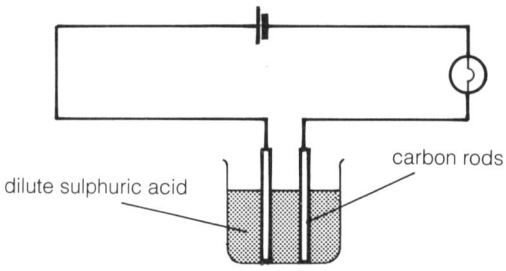

Investigating electrolysis of compounds

Cue Sheet 3:   step 8

substance	bulb	anode	cathode
hydrochloric acid calcium hydroxide sodium bromide potassium iodide copper sulphate			

The anode is the *positive* electrode.
The cathode is the *negative* electrode.